GLOSSARY
of
BIOCHEMISTRY
and
MOLECULAR
BIOLOGY

Revised Edition

This Glossary is dedicated to my former students whose needs initiated the effort, to Hermona Soreq and her Hebrew University research team who have been most helpful in guiding me through some of the arcana of molecular biology, to Sarah Bell and her panel of Portland Press reviewers whose constructive criticism materially advanced the work, and to my wife Jacqueline whose patience and encouragement made it possible.

GLOSSARY
of
BIOCHEMISTRY
and
MOLECULAR BIOLOGY

Revised Edition

David M. Glick

PORTLAND PRESS
London and Miami

Published by Portland Press Ltd, 59 Portland Place, London W1N 3AJ, U.K.

In North America orders should be sent to Ashgate Publishing Co., Old Post Road, Brookfield, VT 05036-9704, U.S.A.

First edition published in 1990 by Raven Press.
This revised edition published in 1997 by Portland Press Ltd.

ISBN 1 85578 088 7

British Library Cataloguing-in-Publication Data

A catalogue record for this book is available from the British Library

All profits from the sale of this publication are returned to the Biochemical Society for the promotion of the molecular life sciences.

Typeset by Portland Press Ltd.

Printed in Great Britain by Information Press Ltd.

Preface

In the sciences, an essential aspect of recognizing, recalling and communicating something, be it a substance, relationship or method, is the naming of it. To create names for new concepts, methods or items, we invent new words, often using the roots of a Classical language (apoptosis, glycocalyx, isosbestic), and we recruit familiar words and invest them with new meanings (chaperone, kringle, library). Nomenclature is no trivial matter: it should be a key to understanding new areas of knowledge, but in its complexity, too often it is a lock. Compounding the problem are the inside jokes: the *cognoscenti* speak in cute acronyms, such as SAAB (selected and amplified protein binding site oligonucleotide method). While fun for the insiders, to the outsider this can be very frustrating. This Glossary lists nearly 3000 terms – some appear only in earlier literature, some are very current, some are common terms invested with new meanings, some are lab jargon; in other words old, new, borrowed and blue – and gives succinct definitions to them. The explanations assume a basic familiarity with the biosciences, and should therefore be useful to advanced students and to workers in related fields who wish to converse with biochemists or molecular biologists in their native tongue. I sincerely hope that this Glossary will help to open doors for its users.

As in the first edition (Raven Press, 1990), names of metabolites, enzymes, etc. are not systematically listed, as this information is easily available in a textbook, and accessed through its index. The 'official' nomenclature of enzymes is available in *Enzyme Nomenclature: Recommendations (1992) of the Nomenclature Committee of the International Union of Biochemistry and Molecular Biology on the Nomenclature and Classification of Enzymes* (Webb, E.C., ed.), Academic Press, San Diego, 1992. The nomenclature of other biochemical substances (metabolites, macromolecules, vitamins, hormones, etc.) is presented in *Biochemical Nomenclature and Related Documents: A Compendium*, 2nd edn., published for the International Union of Biochemistry and Molecular Biology by Portland Press, London, 1992.

Notes on the use of this Glossary

In many instances, several items are discussed together, because defining each of them seperately would be both cumbersome and repetitive. In some entries, therefore, the reader is directed to '*see*' another entry. Where one term is synonymous with another (or is an abbreviation), the defined entry is designated by '='. In addition, where an entry contains terms that are defined elsewhere these are italicized.

This revised edition of the Glossary introduces several changes. Most importantly, references are included with many of the entries in order to assist the reader in searching for additional information. These citations, frequently review articles, are by no means intended to indicate an original source; they are offered merely as convenient entrées into the literature on the subject. Another innovation is a bibliography of additional references.

Readers are invited to contribute suggested entries, particularly with appropriate literature citations, for hoped-for future editions. The collection of new terms for an expanding area – perhaps 'exploding' is a better word – is a difficult task, and any help would be much appreciated. Please send suggestions for new terms to edit@portlandpress.co.uk

David Glick
The Hebrew University of Jerusalem
April 1996

A

(A + T)/(G + C) ratio A reference to the base composition of double-stranded DNA. DNA from different sources has different ratios of the A-to-T and G-to-C base pairs, e.g. DNAs isolated from organisms that live in hot springs have a higher GC content, which takes advantage of the increased thermal stability of the GC base pair. (*see also* Chargaff's rule)

ab initio Completely from the beginning, as opposed to pre-existing, e.g. an *ab initio* complex is one that is assembled from its completely separated subunits, or an *ab initio* calculation is one that is constructed from theory, rather than from an empirical correlation.

absolute configuration The actual, as opposed to relative to some other compound, orientation of atoms in space at an asymmetrical centre.

absorbance (*see* Beer–Lambert equation)

absorption In immunology, the neutralization of immune serum by some of the antigens against which it has been raised. Serum raised against a hapten attached to a larger protein may be exposed to the carrier protein alone in order to absorb the antibodies directed towards epitopes of the protein, leaving the serum with un-neutralized antibodies directed only towards the hapten.

absorption spectrum The molar absorption (extinction) coefficient as a function of wavelength, usually displayed with absorbance on the ordinate and wavelength on the abscissa.

Abu shunt An alternative pathway in bacteria and higher plants for entry of glutamate into the tricarboxylic acid cycle: glutamate → γ-aminobutyrate (Abu) → succinic acid semialdehyde → succinate.

abzyme (= catalytic antibody)

accelerated diffusion (= facilitated diffusion)

acceptor In enzyme mechanisms, a functional group of an enzyme that transiently receives a moiety of a substrate, the *donor*, before itself becoming a donor in transferring it to a second substrate, which is also an 'acceptor'. More generally in immunology, pharmacology and cell biology, an acceptor is an entity that receives an atom, ligand or structure from a 'donor'. (*see also* class II receptor)

acceptor (3′) splice site A site in pre-mRNA that corresponds to the 3′-end of the intron and the 5′-end of the next exon. The last two bases of the intron at the acceptor splice site are AG in most cases, but AC in some instances.

• Mount, S.M. (1996) Science **271**, 1690–1692

acceptor stem (*see* cloverleaf)

acetal (*see* hemiacetal)

acetate rule The observation, first based on inspection of structures, later on experimentation, that many natural products appear to have been assembled from multiple acetate (acetyl-CoA and/or malonyl-CoA) units in head-to-tail condensations. (*see also* acetogenin; polyketide; propionate rule)

acetogenin A compound derived from acetyl units donated by acetyl- and/or malonyl-CoA units, assembled into a non-reduced polyketide, i.e. with the carbonyl groups intact, often, then, cross-linked by aldol condensation and processed by further biochemical transformations to the final product, e.g. orsellinic acid, griseofulvin. (*see also* acetate rule; depside; polyketide)

• Katz, L. and Donadio, S. (1993) Annu. Rev. Microbiol. **47**, 875–912

Achilles heel cleavage A method for cleaving of DNA at very specific sequences. The site of cleavage is able to bind a specific protein, e.g. the *lac* repressor, that already contains a restriction site or is engineered to contain one; the presence of the DNA-binding protein protects the restriction site while the rest of the DNA is enzymically methylated, thus eliminating all unprotected restriction sites. Upon removal of the binding protein, the DNA can be cleaved at the one remaining restriction site.

• Roberts, L. (1990) Science **249**, 127

acid blob A feature of DNA-binding proteins that activate yeast transcription. The only characteristic of these activating regions seems to be the high density of acidic amino acid residues.

acid protease (= aspartate proteinase)

acinar cell A secretory cell within an acinus.

acinus A cluster of secretory cells surrounding a duct.

acrosome A specialized lysosome of a spermatozoon that contains hyaluronidase, the proteinase acrosin and other hydrolytic enzymes.

ACS Anomalously replicating consensus sequence. (*see* origin recognition complex)

action spectrum A measure of the effectiveness of various wavelengths in promoting photosynthesis; usually a plot of photosynthetic efficiency against wavelength.

activation energy The energy needed to raise the reactants, or an enzyme–substrate complex, to the transition state, where it has an equal likelihood of conversion to product or reversion to reactants. This value is commonly evaluated in an *Arrhenius plot*, $\ln k$ against $1/T$, where k is the rate constant, T is the absolute temperature and the slope is E_a/R (E_a is the activation energy and R the gas constant). (*see also* Q_{10}; reaction co-ordinate)

activator In enzyme kinetics, a compound that increases the rate of an enzymic reaction. (*see also* allostery)

active acetaldehyde The CH_3CHOH-group when attached to the thiamin pyrophosphate prosthetic group of pyruvate decarboxylase in the pyruvate dehydrogenase complex.

active glycolaldehyde The α-hydroxy-ethylidene group when attached to the thiamin pyrophosphate prosthetic group of transketolase.

active site The binding and catalytic sites of an enzyme; more loosely, those residues of an enzyme that interact with a substrate or participate in any way in binding or catalysis.

active transport An energy-requiring transport mechanism; one that works against a concentration gradient. (*see also* facilitated diffusion; passive diffusion)

acyl-enzyme An intermediate in the hydrolysis of substrates by some peptidases and esterases, e.g. by serine proteinases, in which the acyl moiety of the substrate is transiently attached to a serine hydroxy group of the enzyme.

acylium ion (*see* a-type ion)

adaptation The evolution of a feature or function through natural selection of incremental improvements, as contrasted with exaptation. (*see also* junk DNA)

adaptor hypothesis The proposal, made before the roles of mRNA and tRNA in protein synthesis were established, that a low-molecular-mass form of RNA provides the interface between each amino acid and the template. The hypothesis assumed that the interaction of the adaptor and a polynucleotide template is much more specific than that between the amino acid and the template.

adenylate energy charge A measure, on a scale of 0–1, of the degree of phosphorylation of adenine nucleotides; $([ATP] + 1/3[ADP])/([ATP] + [ADP] + [AMP])$. (*see also* phosphorylation potential)

adipo- A prefix that indicates fatty tissue, e.g. adipocyte (a fat cell).

adipose tissue Fatty tissue.

A-DNA A right-handed helix; a variant of the dominant B-form of DNA in solution, in which the base pairs are tilted somewhat out of the perpendicular orientation to the axis of the helix. (*see also* Z-DNA)

adrenergic Responsive to the adrenal medullary hormone, i.e. adrenaline (epinephrine), and by extension responsive to other catecholamines. β-Adrenergic responses are those that result in the intracellular generation of cyclic AMP.

affinity chromatography The separation of soluble macromolecules by use of a stationary phase that is designed to interact specifically with, and thus retard the elution of, the desired material; e.g. a hapten attached to a resin to help isolate an immunoglobulin directed against it.

affinity cleavage A variation of the footprinting technique in which a DNA molecule is cleaved at a site occupied by a binding protein. A metal chelator [e.g. ethylenediaminetetra-acetate (EDTA)] is covalently attached to an amino acid residue of the binding protein; Fe^{3+} is ligated to the modified binding-protein–DNA complex and reduced *in situ* to generate free radicals that cleave the DNA in the immediate vicinity. A DNA sequencing gel ladder subsequently indicates the site of binding.

• Oakley, M.G. and Dervan, P.B. (1990) Science **248**, 847–850

affinity labelling A technique that depends upon the tight attachment of a ligand to a binding site on a protein or cell, followed by a chemical reaction to link the ligand covalently to its binding site. (*see also* photoaffinity labelling)

affinity precipitation A technique for purification of proteins that depends upon reversible attachment to a ligand. A *bis-ligand* (or *homofunctional ligand*) is a double-headed molecule that can attach at each end to part of a multisubunit protein, forming an insoluble lattice. In an alternative approach, a ligand attaches a protein to a water-soluble polymer that can be made insoluble by changing the conditions of pH, temperature or ionic strength.

• Irwin, J.A. and Tipton, K.F. (1995) Essays Biochem. **29**, 137–156

AFM (= atomic force microscopy)

agarose A highly purified agar derivative that is used as an electrophoresis and chromatography support.

agglutination The clumping together of cells that are suspended in a fluid.

agglutinin A compound that cross-links cells, e.g. a lectin, an antigen that reacts with sensitized cells.

aggregation The formation of higher-molecular-mass species due to non-covalent adherence of smaller species. Especially for proteins, aggregation is a form of denaturation in which non-polar surfaces of secondary structures, e.g. those of α-helices and β-sheets that normally form intramolecular interactions and are buried within the interior of the protein, are allowed to interact intermolecularly and to form multimolecular forms that are sometimes insoluble. Aggregation is contrasted with *oligomerization*, the normal interaction of native, correctly folded, proteins into higher-order multimers.

aggretope A consensus sequence that permits a protein, usually a foreign protein, to bind to a particular major histocompatibility complex protein of a T-cell.

aglycone The non-sugar moiety of a glycoside.

agonist A compound, often a hormone or its analogue, that binds to a receptor and elicits a response. (*see also* antagonist)

ala scan A series of structural variants of a peptide synthesized in order to evaluate the contribution of each residue to the peptide's binding. In each variant a different residue is replaced with an alanine residue, and the affinity of the variant peptide is measured and compared with that of the original.

albumin Originally, a protein that is soluble in salt-free water and that will coagulate when heated; also the principal protein of plasma or serum. (*see also* globulin)

aldimine form (*see* quinimine form)

aldol condensation The reversible formation of a bond between two carbons, one of a carbonyl group and the other adjacent to a second carbonyl group; by extension, a condensation of compounds that share that kind of chemistry, e.g. the citrate synthase reaction.

aliquot A representative sample of a fixed proportion. Thus $50\mu l$ from a 25ml solution represents $1/_{500}$ of the total.

alkaline Bohr effect (*see* Bohr effect)

alkaloid A nitrogen-containing natural product of a plant, often with pharmacological properties, e.g. morphine, nicotine, strychnine.

allele One member of a pair of homologous genes in a diploid cell. An individual with identical alleles at a genetic locus is a *homozygote*; one with non-identical alleles is a *heterozygote*. In a case in which one allele leads to an observable gene product and the other has no phenotype, the functional allele is said to be *dominant* and the non-functional allele *recessive*.

allele-specific PCR (ASPCR) The adaptation of a PCR primer to selectivity for only one variant sequence. The 3'-end of the primer is designed to be complementary to only one allelic sequence, so the appearance of a PCR fragment is indicative of the presence of the variant sequence.

- Ugozzoli, L. and Wallace, R.B. (1991) Methods Companion Methods Enzymol. **2**, 42–48

all-or-none assay A technique to measure the total amount of a functional enzyme, regardless of its efficiency or affinity for its substrate.

allosteric effector A compound that modifies the activity of an enzyme, or its affinity for its substrate, by binding to a site distinct from the active site; a positive effector increases the activity, a negative effector decreases the activity. (*see also* heterotropic enzyme; homotropic enzyme)

allostery Allosteric regulation; the modification of binding or catalytic properties of a protein by binding of a regulator at a site distinct from the ligand- or substrate-binding site. Allostery typically results in sigmoid kinetics. (*see also* heterotropic enzyme; homotropic enzyme)

allotype A classification of immunoglobulin molecules according to the antigenicity of the constant regions; a variation that is determined by a single allele. (*see also* idiotype; isotype)

alpha/beta (α/β)-barrel A structure common to a large group of proteins, all of which are enzymes, defined by their similar structural motif. A ring of eight parallel β-strands makes up the core; the end of each β-strand is connected by a loop to an α-helix that brings the polypeptide chain to the beginning of the adjacent β-strand. (*see also* beta-barrel)

alpha/beta (α/β)-hydrolase fold A structure common to a group of proteins, including esterases, carboxypeptidases and dehalogenases, that have a catalytic triad of nucleophile, histidine and carboxylate positioned on a scaffolding of eight β-sheets connected by α-helices. (*see also* topology/packing diagram)

- Ollis, D.L., Cheah, E., Cygler, M., et al. (1992) Protein Eng. **5**, 197–211

alpha (α)-cell (*see* islet)

alpha (α)-configuration In steroid chemistry, the orientation of substituents below the plane of the ring system, i.e. on the side opposite the angular methyl groups at C-10 and C-13, which have the β-configuration. Adjacent steroid rings may be *trans*-fused (the non-ring substituents are on opposite sides of the plane of the molecule) or *cis*-fused (the substituents are on the same side).

alpha$_1$ (α_1)-cysteine proteinase inhibitor An enzyme inhibitor that also functions as a *kininogen*; specifically H-kininogen and, in the rat, also T-kininogen. (*see also* cystatin)

alpha$_2$ (α_2)-cysteine proteinase inhibitor (= L-kininogen; *see* kininogen)

alpha (α)-helix A secondary structure in proteins; the right-handed helical folding of a polypeptide such that amide nitrogens share their hydrogen atoms with the carbonyl oxygens of the fourth amide bonds towards the C-terminal end of the polymer.

alpha (α)-isomer In sugar chemistry, the anomer that places the hemiacetal (or hemiketal) hydroxy group on the side of the pyranose (or furanose) ring opposite the non-ring carbon atom (i.e. C-6 of glucose) that is attached to the carbon whose configuration defines the sugar as having a D- or L-configuration, i.e. C-5 of glucose; thus the mirror image of α-D-glucose is α-L-glucose. For example, for the pyranose form of α-D-glucose, the C-1 hydroxy group is on the opposite side of the ring from C-6. (*see also* Haworth projection)

alpha (α)-oxidative decarboxylation (*see* oxidative decarboxylation)

altered-self hypothesis The proposal that helper T-cells recognize a foreign material, e.g. a viral antigen, when it is presented on the surface of an antigen-presenting cell as a complex with the major histocompatibility complex (MHC) class II glycoprotein. This is contrasted with a less favoured explanation, the *intimacy* or *dual-recognition model*, which postulates that helper T-cells must simultaneously recognize both the foreign antigen and the separate MHC complex. By a simple extension, the altered-self hypothesis further proposes that a chemically altered MHC may also be recognized by helper T-cells.

alternative splicing The regulation of protein production at the level of mRNA processing. Control of the sites of excision of introns can switch between expression of different proteins, or can regulate the level of expression of a single protein product by producing a translatable or untranslatable mRNA.

• Maniatis, T. (1991) Science **251**, 33–34

***Alu*-PCR** (*see* interspersed repetitive sequence-PCR)

***Alu* sequence** A small interspersed repeat element; a repetitive sequence in human DNA found in some introns, and characterized by *Alu*I restriction sites; the locus of some homologous recombinations. Comprising 3–6% of the genome, the sequences are about 300 base pairs long and usually feature a head-to-tail tandem repeat.

amber mutation A nonsense mutation; the formation of a non-functional protein due to the premature appearance in mRNA of the terminator codon UAG. (*see also* ocher mutation; opal mutation)

ambidexteran An aggregate of glycosaminoglycans that, unless prevented by steric hindrance or over- or under-sulphation, presents one or both surfaces of their tape-like shape to the complementary surface of another; this results in linear polymers if aggregation is on only one surface, or sheets if aggregation can occur on both surfaces.

• Scott, J.E. (1992) FASEB J. **6**, 2639–2645

ambo A prefix denoting a (not necessarily equal) mixture of D- and L-isomers, e.g. in an amino acid sequence, Ala-*ambo*-Glu-Gly = Ala-Glu-Gly + Ala-D-Glu-Gly.

Ames test (*Salmonella* test) A test for mutagenicity and carcinogenicity which uses specially constructed bacterial strains, by screening the effects of test compounds for their ability to produce reverse mutations that will restore the ability of the bacteria to grow in the absence of an essential metabolite.

aminimide A peptide analogue in which an amino nitrogen substitutes the α-amino group of an amino acid residue; this forms a stable, soluble product.

amino acid Usually an α-amino acid, in which a carboxy and an amino (or imino) group are attached to the α-carbon; triple- and single-letter codes are shown:

alanine	Ala	A
arginine	Arg	R
asparagine	Asn	N

aspartic acid	Asp	D
cysteine	Cys	C
glutamic acid	Glu	E
glutamine	Gln	Q
glycine	Gly	G
histidine	His	H
isoleucine	Ile	I
leucine	Leu	L
lysine	Lys	K
methionine	Met	M
phenylalanine	Phe	F
proline	Pro	P
serine	Ser	S
threonine	Thr	T
tryptophan	Trp	W
tyrosine	Tyr	Y
valine	Val	V
aspartic acid or asparagine	Asx	B
any residue that upon acid hydrolysis yields glutamic acid (Glu, Gln, Glp, Gla)	Glx	Z
4-carboxyglutamic acid	Gla	
hydroxyproline	Hyp	
pyroglutamic acid	Pgl, <Glu, Glp	
unknown or unspecified	Xaa	

amino acyl site The part of a ribosome that binds one amino acyl-tRNA where it will accept the peptidyl group held at the peptide site in the form of its tRNA ester. (*see also* peptide site)

ammonotelic Descriptive of an organism that excretes ammonia as an end-product of nitrogen metabolism. (*see also* ureotelic; uricotelic)

amniocentesis A procedure to obtain fetal cells during pregnancy by puncturing the womb with a needle and removing some fluid that surrounds the fetus. The procedure is used, in part, to obtain fetal cells from which DNA can be isolated for pre-natal genetic analysis.

amphibolic pathway A metabolic pathway that participates in both anabolic and catabolic pathways, e.g. the tricarboxylic acid cycle.

amphipathic Having both polar and non-polar groups, e.g. a detergent.

amphipathic helix A protein structure that serves in part as an interface between polar and non-polar phases; an α-helix that displays non-polar residues on one side and polar residues on the other (e.g. in many globular proteins).

amphitropic Having an affinity for both lipid and aqueous environments, e.g. a membrane-associated protein that has domains that are embedded in the membrane and others that extend into the cytoplasm or the extracellular space.

ampholyte A molecule with both an acidic and a basic group, e.g. an amino acid. (*see also* polyampholyte)

amphoteric Having both acidic and basic groups, e.g. an amino acid.

amplicon A cloned, amplified (by PCR), DNA sequence. (*see also* polymerase chain reaction; representational difference analysis)

amplification The causation of a quantitatively major biochemical or physiological event by a quantitatively minor initiator, e.g. the blood clotting cascade, ion gating; also, using the polymerase chain reaction, the increase by orders of magnitude of copies of a DNA template.

amplimer A PCR primer. (*see also* polymerase chain reaction)

anabolism Those energy-requiring metabolic pathways that result in synthesis of macromolecules and their building blocks, e.g. gluconeogenesis, fatty acid synthesis. (*see also* amphibolic pathway; catabolism)

anabolite A metabolite built up from more simple compounds. (*see also* catabolite)

anaerobic glycolysis (*see* glycolysis)

analyte Material evaluated in an assay.

analytical ultracentrifugation A technique of very-high-speed centrifugation that sediments soluble macromolecules and characterizes them according to their rate of sedimentation (*sedimentation-velocity ultracentrifugation*) or the extent of their sedimentation (*equilibrium sedimentation ultracentrifugation*). (*see also* density-gradient centrifugation)

anaplerotic pathway Metabolic reactions that replenish the pools of intermediates of the tricarboxylic acid cycle. These pools may become depleted, as they also serve as precursors for amino acid synthesis, gluconeogenesis and other anabolic reactions.

anchored PCR A variation of the PCR technique, similar to ligation-mediated PCR, that is applied to double-stranded DNA fragments for which the sequence at only one end of the gene is known. The technique allows amplification of a complete sequence of a gene when only the N-terminal sequence of a protein is known. A short polynucleotide of known sequence is ligated to the 3'-ends of the double-stranded DNA so that a primer complementary to it can be added, along with the primer determined from the partial protein sequence. (*see also* ligand-mediated PCR)

• Appenzeller, T. (1990) Science **247**, 1030–1032; Loh, E. (1991) Methods Companion Methods Enzymol. **2**, 11–19

anchored reference marker One of many genes that are used in mapping of chromosomes. Having a well defined locus within a chromosome, it is used to test for genetic linkage in order to localize new traits; also, in comparative genetics, to test for conservation of linkages which may be of functional significance.

• O'Brien, S.J., Womak, J.E., Lyons, A., et al. (1993) Nature Genet. **3**, 103–112

androgen A compound, usually a steroid, that supports the development of male secondary sex characteristics, e.g. testosterone.

anergy In cell biology, inactivation; e.g. clonal anergy, the inactivation of a lymphocyte, as opposed to clonal proliferation or clonal deletion.

• Koshland, D.E., Jr. (1990) Science **248**, 1273

anhydro- A prefix signifying a product of the removal of the elements of water.

anion exchanger (*see* ion-exchange chromatography)

anisotropic Descriptive of a physical property that varies with the angle of observation. (*see also* isotropic)

anisotropy A measure of the mobility of a fluorophore: $A = (I_\parallel - I_\perp)/(I_\parallel + 2I_\perp)$, where I is the intensity of emission and $\parallel$ and $\perp$ indicate polarization parallel and perpendicular respectively to the exciting light. A mobile fluorophore is able to reorientate itself within its fluorescence lifetime and therefore emits unpolarized light; an immobile fluorophore does not reorientate itself, and hence it emits light polarized in the plane of the excited light. (*see also* polarization)

annealing (hybridization) The time- and temperature-dependent process by which two complementary single-stranded polynucleotides associate to form a double helix.

anomalously replicating sequence (*see* origin recognition complex)

anomalous scattering (resonance scattering) A property of diffracted radiation that is useful in X-ray crystallographic analysis for identification of diffracting centres. The otherwise symmetrical diffraction pattern is subtly perturbed when incident energy of an appropriate wavelength interacts with a diffraction centre (in practice, a sulphur or heavy metal atom) that changes the phase of the diffracted radiation in a manner that is characteristic of that atom.

anomer One of two possible compounds that arise when the open-chain form of a sugar condenses via a hemiacetal or hemiketal bond and produces a new asymmetrical centre. (*see also* alpha-isomer; beta-isomer)

antagonist A compound, often an analogue of a hormone, that binds to a receptor but elicits no response. (*see also* agonist)

anthrogenic Descriptive of a cell in which all the chromosomes are of paternal origin; as opposed to *gynogenic*, descriptive of a cell in which all the chromosomes are of maternal origin.

anthrone reaction A colorimetric method for estimation of sugars that involves treatment with sulphuric acid and anthrone.

antibiotic A natural, synthetic or semi-synthetic product, especially a pharmaceutical, that inhibits bacterial growth.

antibody An immunoglobulin molecule that reacts specifically with another (usually foreign) molecule, the *antigen*.

antichimaeric assistance (*see* entropy effect)

anticodon The three-nucleotide sequence of a tRNA molecule that is complementary to a triplet of mRNA (the *codon*) which specifies a certain amino acid.

anticodon arm (*see* cloverleaf)

anti **conformation** In nucleic acid chemistry, the orientation about the glycosidic bond of a nucleoside or nucleotide that places the base away from the sugar moiety; contrasted with the *syn conformation*, in which the base and sugar are oriented towards each other. (*see also* Z-DNA)

anti-ergotypic Descriptive of killer T-cells that recognize and respond to a cell-surface marker on a T-cell that is actively secreting immunoglobulin; contrasted with *anti-idiotypic*, descriptive of killer T-cells that are more restricted in that they recognize and respond to a cell-surface marker on T-cells that produce a specific immunoglobulin idiotype.

antifolate An antimetabolite that blocks the action of tetrahydrofolic acid-dependent reactions, usually by inhibition of folic acid reductase.

antigen (immunogen) A substance that causes production of an antibody directed against itself.

antigenic determinant (= epitope)

antigenized antibody A molecularly engineered antibody that incorporates epitopes of a non-antibody antigen in the complementarity determining regions of the heavy- and light-chain V domains.

• Zanetti, M. (1992) Nature (London) **355**, 476–477

antigen presentation The appearance on the surface of a cell of a foreign protein, e.g. one that arises from viral infection, in a complex with a class I major histocompatibility complex protein. It is this complex that is recognized by killer T-cells.

• Barinaga, M. (1990) Science **250**, 1657–1658

anti-idiotypic (*see* anti-ergotypic)

anti-Lepore haemoglobin (*see* Lepore haemoglobin)

antimetabolite An inhibitor of a key enzyme in metabolism, used to suppress the activity of the cell; often used in chemotherapy.

anti-mutator DNA polymerase A polymerase with a higher than usual degree of fidelity in proofreading.

• Reha-Kranz, L.J. (1995) Trends Biochem. Sci. **20**, 136–140

anti-oncogene (tumour suppressor gene) A gene that normally functions to regulate cell proliferation by suppression of the function of an oncogene; one cause of cancer is the loss or damage of anti-oncogenes.

• Klein, G. (1993) FASEB J. **7**, 821–825

antioxidant A scavenger of reactive oxygen free radical species ($OH^•$, $O_2^{-•}$, etc.) and other oxidizing compounds; of particular importance are those that act as antioxidants *in vivo*, e.g. ascorbate, α-tocopherol, reduced glutathione. (*see also* oxidative stress)

antiparallel In protein chemistry, the orientation of extended polypeptide chains that interact in a pleated sheet structure, one chain in an N- to C-terminal direction and the other in a C- to N-terminal direction; in nucleic acid chemistry, the orientation of the two polynucleotide chains of a double helix, one that runs in a 3' to 5' direction and the other in a 5' to 3' direction.

antiport A transport mechanism that simultaneously drives two different compounds or ions in opposite directions across a membrane. (*see also* mobile barrier; mobile carrier; symport; uniport)

antisense Descriptive of an endogenous or semi-synthetically produced oligoribonucleotide complementary to mRNA and capable of base-pairing and annealing with mRNA to prevent translation; or of an oligodeoxyribonucleotide capable of binding to the major groove of polypurine–polypyrimidine sequences of DNA by Hoogsteen base pairing to silence a gene. Also used to describe one of the two strands of double-stranded DNA, usually that which has the same sequence as the mRNA, i.e. the non-transcribed strand. However, there is not universal agreement on this convention, and a preferred designation is *coding strand* for the strand whose sequence matches that of the mRNA, and *non-coding strand* for the complementary strand (i.e. the transcription template, or transcribed strand).

• De Mesmaeker, A., Haner, R., Martin, P. and Moser, H.E. (1995) Acc. Chem. Res. **28**, 366–374; Hengen, P.N. (1996) Trends Biochem. Sci. **21**, 153–154; Cornish-Bowden, A. (1996) Trends Biochem. Sci. **21**, 155

antisense drug (= code blocker)

anti-terminator A bacteriophage protein that prevents the normal termination of transcription, e.g. the N protein that binds to nut (N utilization) sites, thus countering the action of the rho protein.

apical (luminal) Descriptive of the free border of an epithelial cell, where it is in contact with vascular space. (*see also* basolateral)

apocrine Descriptive of a secretion mechanism in which vesicles that contain the product burst through the cell membrane and are released along with some of the cytoplasm and plasma membrane. (*see also* exocytosis; holocrine)

apoprotein A protein stripped of any prosthetic group or metal ion normally associated with it. (*see also* holoprotein)

apoptosis A morphologically characterized process of programmed cell death, initiated by various physiological or pathological causes (e.g. cell turnover, hormone-induced atrophy, cell-mediated immune cytolysis, tumour regression), that is characterized by shrinkage of the nucleus and cytoplasm, cell fragmentation and phagocytosis. Apoptosis is contrasted with *necrosis*, which is a random pathological process initiated by irreversible cell damage. Apoptosis is controlled by extracellular signals or the removal of extracellular suppressors of cell death. (*see also* death gene)

• Raff, M.C. (1992) Nature (London) **356**, 397–400; Martin, S.J., Green, D.R. and Cotter, T.G. (1994) Trends Biochem. Sci. **19**, 26–31; Ffrench-Constant, C. (1992) Curr. Biol. **2**, 577–579; Kroemer, G., Petit, P., Zamzami, N., et al. (1995) FASEB J. **9**, 1277–1287

apparent K_m The Michaelis constant as observed under conditions (e.g. the presence of a competitive inhibitor) that would hinder the determination of its true value; in the case of a two-substrate enzyme, the Michaelis constant measured under the particular conditions of a defined concentration of the invariant substrate.

approximation (*see* entropy effect)

aprotinin Bovine pancreatic proteinase inhibitor.

AP site A position in a double-stranded DNA sequence that is missing an A or G base (apurinic) or a C or T base (apyrimidinic).

aptamer A polynucleotide ligand that is selected for its ability to fit a target protein or other polymer. (*see also* shape library)

apurinic DNA A polynucleotide that has lost one or more purine bases due to the lability in acid of the glycosidic bond to purines.

arbitrary primer (*see* primer).

Archaea (*see* eukaryote)

archaebacteria (*see* eukaryote)

area detector An 'electronic film'; a device for measurement of diffracted radiation, e.g. in X-ray crystallographic analysis; detector devices are densely arrayed over a screen and report to a computer the intensity of incident light and its position on the screen. (*see also* image plate; multiwire data collector)

arginine fork A specific kind of interaction between an RNA and an RNA-binding protein in which the two equivalent guanidinium nitrogens of an arginine residue interact with adjacent phosphates of a non-double-stranded region of the polynucleotide.

- Calnan, B.J., Tidor, B., Biancalana, S., et al. (1991) Science **252**, 1167–1171

Arrhenius plot (*see* activation energy)

ARS Anomalously replicating sequence. (*see* origin recognition complex)

ascus A spore-like form through which yeast cells pass during their life cycle; an outer wall surrounds a diploid cell that undergoes meiosis to form four haploid cells (ascospores) which eventually rupture to yield two diploid daughters.

AS-ODN Antisense oligodeoxyribonucleotide. (*see* antisense)

aspartate proteinase A type of peptidase that has at its active site two aspartate residues. (*see also* cysteine proteinase; metalloproteinase; serine proteinase)

ASPCR (= allele-specific PCR)

association constant (K_a) Reciprocal of K_d (*see also* pK_a)

asymmetrical PCR A protocol for generation of single-stranded DNA. Unequal amounts of primers are used, so that the first PCR cycles generate equal amounts of each strand of the template but later cycles, which have no more of one of the primers, create only one new strand.

- McCabe, P.C. (1990) in PCR Protocols: A Guide to Methods and Applications (Innis, M.A., Gelfand, D.H., Sninsky, J.J. and White, T.J., eds.), pp. 76–83, Academic Press, San Diego

asymmetrical reaction The unequal handling of like groups in a prochiral compound. (*see also meso*-carbon; Ogston hypothesis)

atomic absorption analysis A technique for the quantification of small amounts of a metal in solution. Monochromatic light, often generated by excitation of the element in question, passes through a flame in which a test solution is vaporized, and the non-absorbed light, corrected for the excited emission of the metal in the flame, is detected by a photomultiplier. (*see also* flame photometry)

atomic force microscopy (AFM) Also known as scanning force microscopy; a method for mapping the surface of microscopic (e.g. cellular) and even submicroscopic (e.g. macromolecular) surfaces. As a sharp tip passes over the surface of an object, contact pressure is kept at a minimal constant value by an electronic feedback loop; the whole deflection of the cantilevered tip is measured, for example by deflection of a reflected laser beam.

- Bustamante, C., Erie, D.A. and Keller, D. (1994) Curr. Opin. Struct. Biol. **4**, 750–760

ATPase An enzyme that hydrolyses ATP; usually the partial activity of an enzyme, or system of enzymes, that uses the energy made available by the hydrolysis of ATP to drive an energetically unfavourable process, e.g. the Na^+/K^+-ATPase of cell membranes.

AT queue A model that accounts for chromosomal banding. The microscopic appearance of chromosomes during mitosis (metaphase) displays *chromomeres* and *interchromomeres*, i.e. *G bands* (Giemsa stain) and *R bands* (Reverse) respectively; the former are characterized as AT-rich and gene-poor and by late replication, and the latter as GC- and gene-rich and by early replication. The DNA is presumably organized into wide spring-like coils from which

100 kb G loops extend parallel to the chromosome axis (G bands), extended DNA sequences from which larger loops extend perpendicular to the chromosome axis (R loops), and AT-rich sequences called *matrix-attachment* or *scaffold-associated regions* that anchor the G and R loop regions and create the characteristically stained bands.

• Gardiner, K. (1995) Curr. Opin. Genet. Dev. **5**, 315–322

atrophy The wasting away of an organ and/or its capabilities. (*see also* hypertrophy)

attenuation The response of the synthesis of bacterial mRNA to the nutritional state of the organism, e.g. the decrease in transcription of the *trp* operon in the presence of tryptophan, which is due to the incomplete transcription of a leader mRNA sequence coded for by the attenuator sequence.

attenuator A polynucleotide sequence that occurs between an operon and its closest structural gene. (*see also* attenuation)

a-type ion In mass spectrometry, an N-terminal fragment of a polypeptide produced by cleavage of the bond between an α-carbon and its neighbouring α-carboxy carbon. A *b-type ion* is an α-N-terminal acylium ion which results from cleavage of an amide bond, and *y-type ions* are C-terminal iminium ions from such a cleavage. (*see also* mass spectrometry)

autocatalysis The activation of a pro-enzyme preparation by that fraction of it that has already been activated.

autocrine Descriptive of a secretion that binds to receptors on the surface of the same cell that produces it. (*see also* endocrine; exocrine; paracrine)

autogenous regulation A phenomenon in which a gene for a single protein is regulated by its own promoter and operator, and constitutes a one-protein operon.

autoimmune disease A failure of tolerance; the reaction of an individual's immune system towards some of the individual's own proteins as if they were foreign proteins, e.g. myasthenia gravis, rheumatoid arthritis.

autolysosome (*see* autophagocytosis)

autonomic nervous system A functional division of the peripheral nervous system that consists of those pathways that are under involuntary control, e.g those that regulate the gastrointestinal tract and glandular function.

autophagic vesicle (*see* autophagocytosis)

autophagocytosis The response of a cell to stress in which it forms an *autophagosome*, an organelle that contains parts of the cell's own cytoplasm. Later this will fuse with a lysosome to form an *autolysosome*; all such inclusions that derive from autophagocytosis are termed *autophagic vesicles.*

autophagosome (*see* autophagocytosis)

autophagy The action of a lysosome to digest materials from its own cell. (*see also* heterophagy)

autoradiography A technique for visualization of radioactivity in histological preparations, paper chromatograms or slab gels from electrophoresis by overlaying the surface with X-ray film and allowing the radiation to form an image on the film.

autosomal recessive Descriptive of a non-sex-linked genetic trait that must be inherited from both parents to be expressed at the phenotypic level.

autotroph A cell that can sustain itself on non-organic nutrients, e.g. a photosynthetic cell.

autotrophic In microbiology, descriptive of organisms able to grow on inorganic material alone. (*see also* heterotrophic; mixotrophic)

auxin A type of plant hormone that affects cell size, e.g. indoleacetic acid.

axial Descriptive of the orientation of a substituent on a six-membered ring that is perpendicular to the plane of the ring; the opposite of *equatorial*, which is the orientation substantially within the plane of the ring. (*see also* chair form)

axial ratio A measure of the asymmetry of a macromolecule, assumed to be an ellipsoid, given by the ratio of the major axis to the minor axis; it is evaluated from physical properties, e.g. hydrodynamic behaviour, light scattering.

axon A long projection of a neuron through which it communicates with other cells.

axoneme The fundamental structural unit of eukaryotic flagella and cilia; it is composed of nine microtubular doublets (in cross-section a microtubule with another fused to it) surrounding two microtubule singlets. (*see also* triplet)

axoplasm The cytoplasm of an axon.

B

BAC (= bacterial artificial chromosome)

backbone In chemistry, the longest continuous chain of atoms bonded to each other, exclusive of all others, that comprise a polymer.

bacteria (*see* eukaryote)

bacterial artificial chromosome (BAC) A cloning vector that can accept up to 350 kb fragments for cloning and sequencing of fragments of the human genome. (*see also* cosmid; yeast artificial chromosome)

• Venter, J.C., Smith, H.O. and Hood, L. (1996) Nature (London) **381**, 364–366

bacteriophage A virus that infects bacteria. Many phage have proved useful in the study of molecular biology and as vectors for the transfer of genetic information between cells. *Lytic phage*, e.g. the T series phage that infect *E. coli* (*coliphages*), invariably lyse a cell following infection; *temperate phage*, e.g. lambda bacteriophage, can also undergo a lytic cycle or can enter a lysogenic cycle, in which the phage DNA is incorporated into that of the host, awaiting a signal that initiates events leading to replication of the virus and lysis of the host cell.

bait region A sequence of an α_2-macroglobulin molecule, and of homologous proteins, that contains scissile peptide bonds for those proteinases that it inhibits. (*see also* hook region)

BAL (= British anti-lewisite)

balanced polymorphism The result of selective pressures for and against a deleterious mutation that permits it to persist in a population. An example is the stable presence in Africa of the sickle cell gene due to the protection against malaria enjoyed by heterozygotes, i.e. those with sickle cell trait.

band centrifugation (= density-gradient centrifugation)

band competition assay (= gel retardation assay; *see* gel shift assay)

band compression An electrophoresis artifact in which DNA fragments differing in length by only one nucleotide are unresolved, sometimes observed through a series of consecutive guanine nucleotides.

barophile (*see* extremophile)

basal metabolic rate (= resting metabolic rate)

base In nucleic acid chemistry, one of the nitrogenous compounds, i.e. purines and pyrimidines, that are incorporated into nucleosides, nucleotides and nucleic acids. The most common bases are adenine, cytosine, guanine, thymine and uracil, abbreviated as A, C, G, T and U respectively.

base equivalency rule (= Chargaff's rule)

base flipping The distortion of a double-stranded DNA structure that disrupts a base pair and redirects one nucleoside of the pair outwards, where it can interact with a DNA-modifying enzyme such as a methyltransferase.

• Roberts, R.J. (1995) Cell **82**, 9–12

basement membrane An extracellular network of fibres and glycoconjugates that underlies and strengthens some tissues; an interface between these tissues and the connective tissue that surrounds them.

base pair In a nucleic acid double helix, a purine and a pyrimidine on different strands that interact by hydrogen bonding, most commonly a GC or AT pair.

base roll Variation in orientation of bases in a DNA double helix that permits some tilting of the bases. (*see also* A-DNA; Z-DNA)

basolateral Descriptive of the border of an epithelial cell that is attached to the basement membrane. (*see also* apical)

basophil A polymorphonuclear leucocyte containing granules that react with a histological stain for basic substances.

bathochromic shift A shift to longer wavelengths.

B-cell (B-lymphocyte) A lymphocyte that can be activated to proliferate and to synthesize and secrete a specific immunoglobulin G by binding of the antigen that is recognized by the immunoglobulin G receptor on the cell.

BDA (= boomerang DNA amplification)

B-DNA A right-handed helix; the dominant conformational variant of DNA in solution, in which the base pairs are stacked nearly perpendicular to the axis of the helix. (*see also* A-DNA; Z-DNA)

Beer–Lambert equation $A = \varepsilon\ c \cdot l$; a quantitative definition of the dependence of the absorbance of monochromatic light, A, on the molar absorption coefficient, ε, the concentration of the chromophore, c, and the length of the light path, l.

Beer's law A quantitative treatment of the absorption of monochromatic light by a solution. The equation $A = \log(I_0/I)$ defines A (the absorbance, which is proportional to the concentration of the chromophore), I_0 (the intensity of the incident light) and I (the intensity of light having passed through 1 cm of the solution). T, the transmission, is defined as $A = -\log T$. (*see also* Beer–Lambert equation)

Bence Jones protein The immunoglobulin light chains that are synthesized in large amounts and are secreted into the urine by multiple myeloma patients.

Benedict's solution (*see* reducing sugar)

beta (β)-barrel A form of supersecondary structure: a structure within a globular protein in which six or more extended polypeptide strands (β-pleated sheets) are arrayed cylindrically to form the staves of a barrel.

beta (β)-bend (= reverse turn)

beta–beta (β–β)-hairpin A form of secondary structure in proteins in which an extended polypeptide chain turns sharply to fold back upon itself and forms an antiparallel β-sheet.

beta (β)-cell (*see* islet)

beta (β)-configuration (*see* alpha-configuration)

beta (β)-isomer In sugar chemistry, the anomer that places the hemiacetal (or hemiketal) hydroxy group on the same side of the pyranose (or furanose) ring as the non-ring carbon atom. (*see also* alpha-isomer)

beta (β)-oxidation The series of enzymic reactions that oxidizes fatty acyl-CoA esters and shortens them by removal of the C-terminal two carbon atoms as acetyl-CoA. More narrowly, it is the oxidation of a compound, such as a fatty acid, at the β-carbon.

beta (β)-oxidative decarboxylation (*see* oxidative decarboxylation)

beta (β)-pleated sheet A form of secondary structure of a protein in which the amide hydrogens of a peptide bond of one extended polypeptide sequence are shared with the carbonyl oxygens of a peptide bond on a second polypeptide sequence. A sheet that often consists of three or more polypeptide sequences is said to be parallel (i.e. both adjacent strands run in the same direction; N- to C-terminal) or antiparallel.

beta (β)-ribbon A DNA-recognition motif; a segment of antiparallel β-structure that can fit into the major groove of a double-stranded polynucleotide and bind to it due to very specific interactions between side chains of the protein and phosphate residues and edges of base pairs of the polynucleotide.

• Kim, S.-H. (1992) Science **255**, 1217–1218

beta (β)-structure (= beta-pleated sheet)

beta (β)-turn (= reverse turn)

bi (*see* enzyme mechanism)

BIA (= biomolecular interaction analysis)

bi-antennary (*see* tetra-antennary)

bidirectional replication Synthesis of DNA that is effected by two replication forks that travel away from a single origin of replication.

bile acid One of the products of cholesterol hydroxylation and side-chain oxidation to the level of a carboxylic acid. The carboxylate is often conjugated through an amide bond to a glycine or a cysteic acid. Excreted into the small intestine from the gall bladder, bile acids act as detergents, and aid lipid absorption.

bile pigment One of the highly coloured products of haem degradation.

bilin A tetrapyrrole pigment chemically related to the bile pigments, e.g. phyco-bilin, a bilin of red or blue–green algae.

bimolecular sheet (= lipid bilayer)

binary system In transgenic research, an approach to control expression of one transgene by a second, each initially established in its own pedigree. By crossing the two lines, doubly transgenic animals are created in which the control may become operational. The gene of interest is regulated by an exogenous ligand acting either as a positive regulator that binds to a repressor or as a negative regulator that binds to a transactivator. The repressor or transactivator are products of the second transgene.

binding protein A circulating protein that carries its ligand from one site in the body to another, e.g. thyroxine-binding protein; also any protein specialized for binding a ligand, e.g. a calcium-binding protein.

binding site That region of the surface of an enzyme (or receptor, or binding or transport protein) that holds the substrate or product (or other ligand).

biogeochemical sulphur cycle (= sulphur cycle)

bio-imprinting The induction of a new catalytic specificity in an enzyme in an organic solvent. An enzyme and a weakly binding non-substrate/non-product (e.g. chymotrypsin and *N*-acetyl-D-tryptophan) are suddenly precipitated together from an aqueous solution by addition of a miscible organic solvent. The complex reverts to the native conformation and behaviour when redissolved in water, but if redissolved in a suitable

organic solvent it will be found to possess a new enzymic activity, e.g. for synthesis of *N*-acetyl-D-tryptophan ethyl ester from *N*-acetyl-D-tryptophan and ethanol. (*see also* molecular imprinting)
• Mosbach, K. (1994) Trends Biochem. Sci. **19**, 9–14

bio-inorganic Descriptive of biochemical materials that include elements, frequently metals, not popularly associated with them, e.g. Fe-S redox proteins, cobalamin, Zn fingers, the Fe-Mo centre of nitrogenase.

bioluminescence The chemiluminescent emission of light by a living thing, e.g. firefly, certain fungi.

biomimetic Descriptive of a compound or process that artificially simulates the function of a natural compound or process, e.g. a catalyst that employs the principles of enzyme specificity and rate acceleration.
• Breslow, R. (1995) Acc. Chem. Res. **28**, 146–153

biomimicry The design of structural materials according to examples from biology, e.g. a polymer strengthened by carbon fibres, inspired by the insect exoskeleton, which is a protein matrix reinforced by chitin fibres.
• Amato, I. (1991) Science **253**, 966–968

biomolecular interaction analysis (BIA) The detection by a biosensor of a specific binding event for quantification of the ligand. One kind of BIA is *surface plasmon resonance*, the observation of polarization of light reflected from a surface coated with one of the binding partners. The change in polarization provides, in real time, a measure of binding of the second partner.
• Szabo A., Stolz, L. and Granzow, R. (1995) Curr. Opin. Struct. Biol. **5**, 699–705.

bioremediation Processes that use the capabilities of micro-organisms to treat waste products that may be environmentally harmful and to render them innocuous.

biosensor A device, especially an electrochemical device, that detects some biological event (e.g. respiration, enzymic activity, binding to an antibody) and converts it into an electrical signal that it reports quantitatively and in real time. (*see also* biomolecular interaction analysis)
• Turner, A.P.F. (1992) Essays Biochem. **27**, 147–159

biosynthesis The process by which a biological structure, especially a relatively simple structure, is formed by a sequence of enzymic reactions that starts from common metabolites or 'synthons', e.g. the synthesis of haem from glycine and succinyl-CoA; the synthesis of steroids from acetyl-CoA. For the assembly of larger structures, such as membranes, via non-covalent interactions, the term *synkinesis* has been coined, with the subunits being termed *synkinons*.
• Fuhrhop, J.-H. and Koening, J. (1994) Membranes and Molecular Assemblies: the Synkinetic Approach, Royal Society of Chemistry, Cambridge

bi-product analogue (collected substrate inhibitor) An inhibitor of a hydrolase that incorporates structural features of both products of catalysis, and thus can bridge the S_1 and S_1' subsites. (*see also* specificity subsite)

bis-ligand (*see* affinity precipitation)

bisnor- A prefix that signifies a product of the removal of two $-CH_2-$ groups.

biuret reaction A colour reaction for the quantification of protein in solution. By analogy with the compound biuret (H_2N-CO-NH-CO-NH_2), the peptide backbone of proteins reacts with alkaline copper solutions to produce a violet colour.

bivalency The property of an immunoglobulin G molecule, and some other immunoglobulins, of having two antigen-binding sites.

blast cell (= stem cell)

bleb A small bud-like projection from the surface of a cell.

blood clotting cascade The sequence of reactions, initiated by exposure of blood to extravascular surfaces, that results in a fibrin clot (*see also* extrinsic pathway; intrinsic pathway)

blood group substances The oligosaccharide moieties of glycoproteins that appear in many biological fluids (saliva, urine, milk) as well as on the surface of erythrocytes. These antigens, upon reaction with specific antibodies, cause agglutination of the cells to which they are attached. Examples are the A, B and O antigens.

blotting The transfer by contact of a macromolecule from a two-dimensional separation medium, e.g. paper or polyacrylamide slab, to another surface of higher affinity, e.g. nitrocellulose or nylon. (*see also* dot blotting; electroblotting; northern blotting; Southern blotting; south-

western blotting; western blotting)

blunt ended (flush ended) Descriptive of the structure of double-stranded DNA in which neither strand of the duplex extends further from the end than the other; often the product of cleavage by a restriction endonuclease. (*see also* sticky ended)

B-lymphocyte (= B-cell)

boat form (*see* chair form)

Bohr effect The decrease in the affinity of haemoglobin for oxygen that occurs when the haemoglobin solution is made more acid above pH 6. The opposite occurs below pH 6; hence the physiological phenomenon is the *alkaline Bohr effect*.

boomerang DNA amplification (BDA) A technique similar to PCR that also uses a heat-resistant polymerase and cycles of polymerization, denaturation and annealing, but which requires only one primer. The source DNA is digested with a restriction endonuclease. A universal adaptor is engineered with self-complementary sections, so that it loops back upon itself and has ends that permit ligation to both strands at each end of the restriction fragment. In the first amplification cycle a primer anneals to an internal site and the polymerase copies the primer-binding strand through the adaptor, back along the second strand and past the site that is complementary to the primer-binding site of the first strand. In subsequent cycles the primer will have two sites for initiation of polymerization.

• Hengen, P.N. (1995) Trends Biochem. Sci. **20**, 372–373

boron analogue A derivative of an enzyme substrate in which >B-OH replaces >C=O of an amide or ester as an approximation to the transition state for its hydrolysis. (*see also* transition-state inhibitor)

bouquet (*see* non-collagen collagen)

bottlebrush (*see* proteoglycan)

bp (= base pair)

BPS (= branch-point sequence)

branch migration (*see* Holliday model)

branch-point A single metabolite that is an intermediate in two or more biosynthetic pathways, e.g. pyruvate (a precursor of acetyl-CoA, alanine and oxaloacetate), chorismic acid (a precursor of phenylalanine, tryptophan and tyrosine).

branch-point sequence (BPS) The sequence near the 3′-end of an intron of nuclear mRNA that contains the adenosine residue which, as the intron is excised, will accept the guanosine residue at the 5′-end of the intron.

• Maniatis, T. (1991) Science **251**, 33–34

branch site In RNA splicing, the site to which the 5′-guanylate at the end of an intron is joined to an adenylate residue within the intron through a 2′,5′-phosphodiester bond to form a 'lariat' intermediate. (*see also* self-splicing)

breakthrough organism A progenote; in early evolution, the first organism able to transmit genetic information, even if with only marginal fidelity.

• Benner, S.A. and Ellington, A.D. (1990) Science **248**, 943–944

British anti-lewisite (BAL) 2,3-Dimercaptopropanol, an antidote for poisoning by arsenite which reacts with reduced lipoic acid.

brown adipose tissue Thermogenic fatty tissue with a high content of relatively uncoupled mitochondria; especially prominent in infants, located around the kidneys and neck.

browning reaction (Maillard reaction) Covalent reactions that occur on heating proteins; the formation of cross-links between side chains and with carbohydrates.

brush border The apical surface of intestinal epithelia, characterized by microvilli that extend into the lumen.

b-type ion (*see* a-type ion)

bubble A widened segment of duplicating double-stranded DNA, visualized by electron microscopy, that is formed by two replication forks that travel away from the common origin of replication. The enlargement, or bubble, is a region in which the strands of the original DNA are separated and paired with newly synthesized DNA.

budding The process of formation of vesicles from a membrane, e.g. the formation of transfer vesicles from the endoplasmic reticulum; also the asexual propagation of a yeast cell by mitotic division of the nucleus and pinching off of a daughter nucleus and some of the cytoplasm.

buffer A proton-binding compound that, in the region of its pK_a, competes with water for added protons and makes the pH relatively insensitive to added acid.

buffer capacity A measure of the ability of a solution to maintain its pH in the face of the addition of acid or alkali; a capacity of 1 when 1 mol of acid (or alkali) is added to 1 litre causes a pH fall (or rise) of 1 pH unit.

buffer value (= buffer capacity)

buoyancy factor In analytical ultracentrifugation, one of the factors that determine a macromolecule's rate of sedimentation; $(1 - \upsilon\rho)$, where υ is the partial specific volume of the macromolecule (expressed as ml/g) and ρ is the density of the solution (expressed as g/ml).

bZIP protein A type of Y-shaped homodimeric DNA-binding protein in which the stem is composed of two intertwined α-helices held together as a leucine zipper; the arms of the Y are the basic regions that reach around the double-stranded DNA and interact with its acidic phosphate moieties in a scissor-like grip. (*see also* helix–turn–helix motif)

C

c A prefix that denotes cellular. (*see also* oncogene)

CAAT box A regulatory sequence upstream from some eukaryotic structural genes.

CAAX box (*see* prenyl)

cafeteria feeding In laboratory animal nutrition, feeding a diet of low nutritional value, i.e. high in calories but low in protein and vitamins.

calcium-induced calcium release (CICR) A mechanism of cellular metabolic control: inositol 1,4,5-trisphosphate, generated in response to cell stimulation, causes a flow of Ca^{2+} across the plasma membrane, which in turn leads to periodic release of Ca^{2+} from the endoplasmic reticulum by an inositol trisphosphate-independent mechanism.

• Berridge, M.J. (1990) J. Biol. Chem. **265**, 9583–9586

calpain A calcium-dependent cellular proteinase with a neutral pH optimum.

calpastatin A protein inhibitor of a calpain.

Calvin cycle (reductive pentose cycle) The series of metabolic reactions by which carbon dioxide is fixed into glycolytic intermediates; the dark reactions of photosynthesis. (*see also* Hill reaction)

canyon hypothesis A proposal to reconcile both the necessary invariability, and apparent non-immunogenicity, of the host cell receptor attachment site of a virus (e.g. a rhinovirus), and the ability of the same virus to mutate into many different immunologically recognized serotypes. The critical attachment site of the virus is sequestered at the bottom of a deep, narrow cleft, a 'canyon', which allows access to the receptor but not to an immunoglobulin; sites at the 'rim' of the canyon are accessible to immunoglobulins, but as they are not critical to the virus, they can mutate freely.

• Rossmann, M.G. (1989) J. Biol. Chem. **264**, 14587–14590

cap The 7-methylguanine nucleoside attached to the 5'-end of mRNA by a 5'-5'-triphosphate bond.

capillary electrophoresis (*see* electrophoresis)

• Landers, J.P. (1993) Trends Biochem. Sci. **18**, 409–414

capping The process of modifying the 5'-end of eukaryotic mRNA with 7-methylguanine. (*see also* cap)

capsid The protein coat of a virus.

cap site The site on a DNA template where transcription begins. It corresponds to the nucleotide at the 5'-end of the RNA transcript which accepts the 7-methylguanine cap.

captive antibody (*see* sandwich immunoassay)

carbanion A transient species in which a carbon atom bears a formal negative charge, e.g. an intermediate in the carboxylation of glutamate residues of preprothrombin. (*see also* carbocation)

carbocation (carbonium ion) Any stable or transitory ion in which a carbon atom bears a formal positive charge. (*see also* carbanion)

carbohydrate One of a class of biological materials comprising sugars, polymers of sugars, and compounds related to them. The name derives from the basic sugar structure, $(CH_2O)_n$. The category includes reduction and oxidation products, phosphate and sulphate esters, and amine derivatives.

carbon cycle The movement of carbon atoms through different chemical forms and locations, from dissolved CO_2 in equilibrium with atmospheric CO_2 through plant carbohydrate, fats and proteins of plants and animals, and via oxidation back to atmospheric CO_2.

carbonium ion (= carbocation)

carboxyl proteinase (= aspartate proteinase)

carboxyl-terminal analysis (= C-terminal analysis)

carcinoma A malignant tumour of mesodermal origin.

cardiac glycoside (see saponin)

cardiac muscle A form of striated muscle characteristic of the heart.

carotenoid A derivative of phytoene, a symmetrical non-cyclic 40-carbon terpene.

carrier Non-isotopically labelled material that dilutes a labelled tracer and enlarges the pool of the labelled metabolite, thus assisting in its isolation.

cartwheel (see triplet)

cascade A series of enzymic reactions that at each step convert an inactive enzyme into an active enzyme, which in turn activates another inactive enzyme, and thus greatly amplifies the initial signal. In a *homocascade* all the events involve one enzymic type (e.g. proteolysis); in a *heterocascade* the events involve different enzymic types (e.g. phospholipase and kinase).

cassette A mutation-containing restriction fragment that can replace the homologous fragment excised from the genome of an organism.

cassette mutagenesis A method to determine the limits of tolerance of a protein to amino acid substitution. A cloned gene is mutated by synthesis of short segments of the gene with random base substitutions, insertion of these altered polynucleotide sequences into the gene and then transformation of cells with the mutated gene. Cells that contain a functional protein are selected and the protein (or gene) is examined to identify the successful mutation. Testing of many such short segments until the entire structural gene is examined maps the areas of tolerance and of sensitivity of the gene product to substitutions.

• Bowie, J.U., Reidhaar-Olsen, J.S., Lim, W.A. and Sauer, R.T. (1990) Science **247**, 1306–1310

CASTing (= cyclic amplification and selection of targets)

catabolism The action of energy-yielding metabolic pathways that degrade macromolecules and complex compounds or small molecules into CO_2, H_2O, etc. (see also amphibolic pathway; anabolism)

catabolite A degradation product derived from a more complex compound. (see also anabolite)

catabolite activation (see glucose effect)

catabolite repression (see glucose effect)

catalytic antibody (abzyme) An antibody with catalytic properties; often raised against a hapten that chemically resembles the transition state of the intended substrate so as to force a substrate into that transition state.

- Hilvert, D. (1994) Curr. Opin. Struct. Biol. **4**, 612–617; Jacobsen, J.R. and Schultz, P.G. (1995) Curr. Opin. Struct. Biol. **5**, 818–824

catalytic-centre activity (*see* turnover number)

catalytic configuration (*see* entropy effect)

catalytic rate constant (k_{cat}) The first-order rate constant that describes the rate-limiting step in enzyme catalysis, usually the conversion of the enzyme–substrate complex into the enzyme–product complex; the maximal velocity divided by the enzyme concentration. (*see also* turnover number)

catalytic site The region of an enzyme that interacts with the substrate to effect the enzymic reaction.

catalytic subunit (*see* regulatory subunit)

catalytic triad (*see* charge relay system)

CAT assay (= chloramphenicol acetyltransferase assay)

catecholamine One of a family of phenolic compounds chemically related to catechol (1,2-dihydroxybenzene), which is derived metabolically from tyrosine; the family comprises hormones and neurotransmitters, including adrenaline (epinephrine), noradrenaline, dopamine, etc.

catenane (= concatenate)

cathepsin An endopeptidase, often with low pH optimum, associated with lysosomes in the cell.

cation exchanger (*see* ion-exchange chromatography)

caveolae Small invaginations of the plasma membrane, characterized by glypiated proteins among others, that provide binding sites and points of entry of small molecules into the cytoplasm via trancy-tosis or potocytosis. *Trancytosis* is the pinching off of the ligand-bearing vesicle into the cytoplasm. *Potocytosis* is the putative closure of the narrow neck of the invagination and consequent isolation of the vesicular space from the extracellular fluid; the high local concentration of the ligand allows diffusion of trapped molecules into the cytoplasm.

- Travis, J. (1993) Science **262**, 1208–1209

ccDNA Closed circular DNA.

C_6C_3 metabolite (= shikimic acid metabolite)

C_3–C_4 photosynthesis A variant of C_3 photosynthesis in which the 2-carbon product of photorespiration is efficiently oxidized to CO_2, which is then recycled in photosynthesis. (*see also* C_4 photosynthesis)

- Rawsthorne, S. (1992) Essays Biochem. **27**, 135–146

C_4 cycle (= C_4 photosynthesis)

CD (= circular dichroism)

cDNA Complementary DNA; DNA that is synthesized, by reverse transcriptase, from an mRNA template, and therefore has no introns. (*see also* genomic DNA)

cDNA library A collection of cells, usually *E. coli*, transformed by DNA vectors each of which contains a different cDNA insert synthesized from a collection of mRNA species. (*see also* genomic library)

CDR (= complementarity-determining region)

cecropin One of a group of inducible antimicrobial oligopeptides found in the haemolymph of a moth that serve some of the same functions as the immune system of higher species. (*see also* magainin)

cell culture An *in vitro* technique for the propagation of genetically homogeneous, dispersed animal or plant cells in a differentiated state, usually in complex growth media. (*see also* explant)

cell cycle The necessary sequence of growth and synthetic stages through which a cell passes, from its origin by mitosis of a parent cell until its own division into daughter cells; consists of a G_1 (first gap) period, an S (synthesis) phase, a G_2 period and an M (mitosis) period.

cell pole One of the two foci of a cell during mitosis, defined by a centriole, from which half the mitotic spindle radiates towards the other pole.

cell-to-cell channel (= gap junction)

cell wall The rigid, mainly cellulose, structure that surrounds the plasma membrane of a plant cell.

centimorgan (cM) A genetic measure of the distance that separates markers in the same chromosome, especially in describing a map of the chromosome; the distance that permits a 1% frequency of crossing over (1 morgan is the distance that permits 100% crossing over); equivalent to about 1×10^6 bases.

central dogma In biology, the proposition that the permanent repository of genetic information is DNA which can be replicated, and that the information is expressed unidirectionally by transcription into RNA and thence by translation into protein; later qualified in important ways, e.g. upon the discovery of reverse transcription.

central nervous system (CNS) A division of the nervous system of higher animals that consists of the brain and spinal cord. (*see also* peripheral nervous system)

centrifugal elutriation A method for separation of isolated cells according to their characteristic sedimentation rates in a centrifuge rotor which is designed to allow flow-through of a fluid during operation; also known as countercurrent elutriation and elutriation centrifugation. The centrifugal force on the cells is opposed by the force of a fluid moving in the opposite direction. Cells are first sedimented in a density gradient, then displaced by a buffer of increasing density that flows into the bottom of the sample cell and out of the top, and carries with it cells of the same density.

• Diamond, R.A. (1991) Methods Companion Methods Enzymol. **2**, 173–182

centriole One of two cylindrical structures that appear in animal cells. During mitosis the centrioles are the poles from which the microtubules of the mitotic spindle radiate; they represent specialized microtubule-organizing centres.

centromere The constriction in a chromosome seen during mitosis; the point of contact with the mitotic spindle from which the chromosome is drawn to a centrosome. Positions along the chromosome are named relative to the centromere: p designates the short arm (that part from the centromere to the nearest end); q designates the long arm. Numbers that follow p or q indicate the number of the G band: thus 7p22 indicates the 22nd band on the short arm of chromosome 7; 3q26-ter indicates a position between band 26 and the end (terminus) of the long arm of chromosome 3. (*see also* AT queue)

centrosome One of the structures of a cell that during mitosis serves as a microtubule-organizing centre; it becomes the structure from which the mitotic spindle radiates and which defines a cell pole.

chair form A conformation of a six-membered non-aromatic ring that places all opposing centres away from each other, as contrasted with the *boat form*, which places two of the opposing centres towards each other. A *twist conformation* is a slightly flattened and relaxed variation of the chair form.

chalone An inhibitory hormone.

channel former A structure in a membrane that, without itself moving, allows passive diffusion of ions across the membrane.

chaotropic agent A solute that disrupts the structure of the bulk water phase and, in so doing, changes the solubility and stability properties of other solutes, such as proteins.

chaperone (*see* chaperone machine; RNA chaperone)

chaperone machine A multicomponent system that ensures the proper folding of nascent proteins or their intracellular transport following synthesis; components include *chaperones* (eukaryotes) or *chaperonins* (from prokaryotes, mitochondria and plasmids), which bind to unfolded proteins; also peptidyl proline isomerases and heat-shock proteins, which can serve various ancillary functions. (*see also* RNA chaperone)

• Burston, S.G. and Clarke, A.R. (1995) Essays Biochem. **29**, 125–136; Hendrick, J.P. and Hartl, F.-U. (1995) FASEB J. **9**, 1559–1569; Ptitsyn, O.B. (1996) FASEB J. **10**, 3–4; Hartl, F.-U. (1996) Nature (London) **381**, 571–580

chaperonin (*see* chaperone machine)

Chargaff's rule An empirical finding that in DNA the frequency of A equals the frequency of T, and the frequency of G equals the frequency of C; later given a theoretical basis by the Watson–Crick double-helix model of DNA.

charge accumulator model A conceptualization of the mechanism of the water-splitting enzyme of photosynthesis in which four electrons are sequentially removed from the manganese centre, enabling it to oxidize water to molecular oxygen.

charge ladder A series of elution peaks seen upon capillary electrophoresis of modified versions of a protein that vary only in their electrostatic charge. (*see also* electrophoresis)

• Gao, J., Gomez, F.A., Harter, R. and Whitesides, G.M. (1994) Proc. Natl. Acad. Sci. U.S.A. **91**, 12027–12030

charge relay system A tautomeric form of a protein in which a formal charge on one dissociating group is moved to another in the same protein. In serine proteinases, for example, an internal β- (sometimes a γ-) carboxylate is partially protonated by bridging via an imidazole to an active-site nucleophile, often a serine hydroxy or cysteine thiol group, leaving the nucleophile with a negative charge and making it more reactive in its attack upon a substrate molecule. This group of residues is termed the *catalytic triad*.

chelator A compound that binds ions, especially metal ions, by several functional groups whose combined effect results in a high-affinity interaction.

chemical cleavage method (= Maxam–Gilbert method)

chemical coupling hypothesis Explanation for oxidative phosphorylation by analogy with substrate-level phosphorylation, i.e. when a phosphate derivative of an electron carrier is oxidized, it becomes a high-energy phosphate compound, energetically capable of transferring its phosphate to ADP. (*see also* chemiosmotic theory; conformational hypothesis)

chemical force microscopy A variant of atomic force microscopy in which the sensing probe is coated with a substance that interacts with the test material or object.

• Frisbe, C.D., Rozsnyai, L.F., Noy, A., et al. (1994) Science **265**, 2071–2074

chemical mismatch detection (= hydroxylamine and osmium tetroxide technique)

chemical shift In nuclear magnetic resonance, the modulation of field strength necessary to achieve the resonance frequency characteristic of a particular atom in a particular chemical, i.e. bonding, environment.

chemiluminescence The light emitted by an exergonic chemical reaction; contrasted with *photoluminescence* (fluorescence or phosphorescence), which is the light emitted as a fluorophore falls to a lower energy state after having been excited by irradiation.

chemiosmotic theory A proposed mechanism of oxidative phosphorylation in the mitochondrion and chloroplast that

requires (1) electron transport to be arranged across the mitochondrial or chloroplast membrane so that protons are vectorially transported to its outer surface, (2) ATP synthesis to be arranged in the membrane so that the proton gradient can be used to drive ATP synthesis, and (3) that the mitochondrial/chloroplast membrane is impermeable to protons and defines an osmotically isolated space. The electron transport chain is composed alternately of hydrogen atom carriers and electron carriers, so that transfer from the former to the latter permits liberation of protons and their vectorial transport across the mitochondrial or chloroplast membrane. Because proton transport is not accompanied by an equivalent transport of electrons across the membrane, it generates not only a chemical potential (ΔpH) but also an electron potential ($\Delta\mu$), and the two together are termed the proton motive force. (*see also* chemical coupling hypothesis; conformational hypothesis; electrochemical gradient; P:O ratio; proton motive force)

chemoattractant A chemotactic agent. (*see also* chemotaxis)

chemotaxis The movement of a cell along the concentration gradient of a chemotactic agent towards its source.

• Stock, A.M. and Mowbray, S.J. (1995) Curr. Opin. Struct. Biol. **5**, 744–751

chemotroph An organism that derives its energy from chemical reactions, usually by the oxidation of nutrients by molecular oxygen. (*see also* phototroph)

chemotype A strain of bacteria defined by variations in a biosynthetic product, e.g.

Gram-negative bacteria chemotypes defined by the chemistry of their lipopolysaccharides.

- Raetz, C.H.R. (1990) Annu. Rev. Biochem. **59**, 129–170

chemzyme A chemical enzyme; a soluble small organic molecule, sometimes with the addition of a transition metal ion, that is capable of catalysing specific reactions.

chiasma The point of attachment of two chromatids during meiosis that results in *crossing-over*, which is the exchange of a region of one chromosome, from an end to the chiasma, with the homologous region of a sister chromosome.

chi-form (*see* Holliday model)

chimaeric antibody (= humanized antibody)

chimaeric DNA A hybrid molecule produced by combining DNA from two different species into a single polynucleotide.

chimaeric protein A protein product of a chimaeric DNA gene.

chimaeric toxin (= immunotoxin)

chip (oligonucleotide array) In DNA technology, a component of a device for screening labelled DNA libraries; a standardized solid surface carrying immobilized DNA probes in a known, specific array, so that automatic recording of absorption, fluorescence, radioactivity, etc. (depending upon the labelling method) may simultaneously score successful hybridizations. More generally, a spatially addressable surface for the synthesis and screening of libraries. (*see also* small molecule library)

- Southern, E.M. (1996) Trends Genet. **12**, 110–115

chirality 'Handedness'; the formal orientation in space of stereoisomers or potentially stereoisomeric compounds. (*see also* CIP classification; *meso*-carbon; Ogston hypothesis; prochirality; *R*; symmetry)

- Cahn, R.S., Ingold, C. and Prelog, V. (1966) Angew. Chem. **78**, 413–447; Dodziuk, H. and Mirowicz, M. (1990) Tetrahedron Asymmetry **1**, 171–186

chloramphenicol acetyltransferase (CAT) assay A procedure for evaluation of the regulatory properties of eukaryotic promoter sequences. The CAT gene (which encodes an enzyme found only in bacteria) is used as a 'reporter gene' in that it is fused to a promoter sequence and introduced into a eukaryotic cell, where the ability of the promoter to cause the expression of the CAT gene is monitored by assay of the enzyme's activity; the assay may involve thin-layer chromatographic analysis of the conversion of [^{14}C]chloramphenicol into acetyl [^{14}C]chloramphenicol.

chloride shift The movement of chloride ions into erythrocytes as bicarbonate exits, due to the generation by carbonic anhydrase of bicarbonate from carbonic acid inside the cells as they pass through peripheral tissues and take up carbon dioxide.

chloroplast An organelle of a green plant cell in which light harvesting and ATP synthesis occur. (*see also* thylakoid membrane)

cholinergic Responsive to acetylcholine.

chondro- A prefix indicating cartilage, e.g. chondrocyte.

Chou–Fasman analysis A method for prediction of the secondary structure of a protein from its primary structure. A survey of proteins of known secondary and tertiary structure allows an empirical classification of amino acid residues according to their abilities to strengthen or destabilize α-helix or β-structure. Recent advances have high-speed computers using Monte Carlo calculations to first fit multi-helix proteins into individual α-helices, then use empirically assigned interaction coefficients to locate the helices in spatial relation to each other.

• Skolnick, J., Kolinski, A., Brooks, C.L., III, et al. (1993) Curr. Biol. **3**, 414–423

chromaffin granule A dichromate-staining vesicle in cells that contains catecholamines or 5-hydroxytryptamine (serotonin); found for example in chromaffin cells of the adrenal medulla.

chromatid One of a pair of chromosomes.

chromatin The complex of DNA and associated proteins, most notably histones, that occurs in the nuclei of eukaryotic cells.

• van Holde, K. and Zlatanova, J. (1995) J. Biol. Chem. **270**, 8373–8376

chromatogram A graphical representation of a chromatographic separation, e.g. absorbance or radioactivity of the eluate (ordinate) plotted as a function of eluate volume (abscissa).

chromatophore An epithelial cell of a lower animal in which pigment granules can be physically moved to effect colour changes.

chromatophoresis A technique for protein separation that uses high-pressure liquid chromatography followed by sodium dodecyl sulphate (SDS)/polyacrylamide-gel electrophoresis. The effluent from a reverse-phase column is mixed with SDS and a reducing agent and applied to a polyacrylamide slab gel; the resulting gel shows two-dimensional separation, by polarity in one dimension and by molecular mass in the other.

chromomere (*see* AT queue)

chromophore A chemical group in a compound or macromolecule responsible for the absorbance of visible or ultraviolet light.

chromosomal banding (*see* AT queue)

chromosomal map The localization of genetic traits to regions of chromosomes, by linkage studies (e.g. the association of haemophilia with the X chromosome) and by *in situ* hybridization to fluorescent probes.

chromosome One of the nuclear structures, composed largely of chromatin, into which eukaryotic genes are organized.

chromosome landing In positional cloning, an alternative to chromosome walking for finding a gene. Clones of genomic DNA are fragmented so as to include both the target gene and a closely linked marker and are screened to select ('land on') those clones that contain the target gene.

• Tanksley, S.D., Ganal, M.W. and Martin, G.B. (1995) Trends Genet. **11**, 63–68

chromosome painting (= coincidence painting)

chromosome walking A strategy for mapping and sequencing a chromosome.

Large restriction fragments are generated and, by Southern blotting with an RNA or cDNA probe, a single starting point is identified. New probes are synthesized from sequences of the same fragment that are adjacent to the starting point, and are then used to identify different restriction fragments adjacent to that which bears the starting point. The procedure is used repetitively, working towards the ends of the chromosome and away from the starting point.

CICR (= calcium-induced calcium release)

cilium A short extension of the plasma membrane with an axoneme core that functions in cell mobility or movement of extracellular substances across the cell surface.

CIP classification The system of nomenclature of asymmetrical compounds devised by Cahn, Ingold and Prelog. (*see also* chirality; *R*)

circadian rhythm The cyclical pattern of daily increases and decreases in enzyme and hormone levels and other physiological functions.

• Kay, S.A. and Millar, A. (1995) Cell **83**, 361–364

Circe effect (*see* entropy effect)

circular dichroism (CD) A technique for determination of the asymmetry of a molecule. Unlike in a conventional polarimeter, where light is restricted to oscillation in a plane, in CD the direction of oscillation turns clockwise around the direction of the beam (right-circularly polarized) or counter-clockwise (left-circularly polarized). CD is expressed as $(A_L - A_R)/lm$, where for any wavelength A_L and A_R are the absorbances of left- and right-circularly polarized light respectively, l is the pathlength through a solution, and m is the molarity of a chiral solute. A CD spectrum of a protein in the far-ultraviolet, where the peptide bond absorbs, is characteristic of its conformation, whether α-helix, β-sheet or random coil. (*see also* optical rotation; optical rotatory dispersion)

• Johnson, W.C., Jr. (1988) Annu. Rev. Biophys. Biophys. Chem. **17**, 145–166

cis In stereochemistry, descriptive of a substituent on the same side of a structure that prevents equalization of positions by restricted rotation, e.g. an olefinic bond, a ring or a peptide bond; contrasted with *trans*, on the same side of the structure. In genetics, *cis* refers to linked markers and *trans* to separated markers.

***cis*-acting** Descriptive of the controlling effect of a regulatory gene on a structural gene that is adjacent to it; later broadened to distinguish intramolecular (*cis*) from intermolecular (*trans*) actions. (*see also trans*-acting)

***cis*-fusion** (*see* alpha-configuration)

***cis*-Golgi** (*see* Golgi apparatus)

***cis*-recognition** (*see trans*-recognition)

cistron A segment of DNA that contains all the information necessary for the production of a single polypeptide and includes both the structural (coding) sequences and regulatory sequences (transcription start and stop signals). (*see also* monocistronic mRNA; operon; polycistronic mRNA)

citric acid cycle (= tricarboxylic acid cycle)

citrovorum factor N^5-Formyltetrahydrofolic acid; originally isolated as a growth factor for *Leuconostoc citrovorum*.

C kinase (= protein kinase C)

clade In phylogenetic taxonomy, families or subfamilies descended from a common ancestor.

class I (*see* major histocompatability complex)

class II (*see* major histocompatability complex)

class II receptor An acceptor; a membrane constituent that aids in transmembrane transport of nutrients, ions, etc.; distinguished from pharmacological receptors, which perform signal transduction, e.g. signal recognition followed by generation of a second messenger.

• Hollenberg, M.D. (1991) FASEB J. **5**, 178–186

-clast A suffix that indicates a cell that destroys or resorbs, e.g. osteoclast.

clathrate A cage-like structure, e.g. that formed by water molecules that surround a hydrocarbon in solution.

cleft The space between domains of a protein, often the binding or catalytic site of an enzyme. (*see also* pocket)

clonal selection theory (= selective theory)

clone In microbiology, a colony of cells all descended from a single ancestral cell and therefore genetically homogeneous; in molecular biology, an exact replica of a DNA fragment; *cloning* is the act of preparation and propagation of a genetically defined cell or a DNA fragment.

cloverleaf The formalized pattern assumed by a tRNA molecule viewed in two dimensions that shows the regions of internal complementarity that allow the polynucleotide to fold back upon itself into base-paired double helices. The stem includes the *acceptor stem* at the 3'- end, which attaches the amino acid, and the non-complementary loops or arms include the *anticodon arm*, which hybridizes with the codon of an mRNA. In three dimensions the structure can be divided into two sections at a right angle to one another: a coaxial stack that includes the acceptor stem, and the remainder of the molecule, i.e. the common arm that is shared by the two sections.

• Pace, N.R. and Smith, D. (1990) J. Biol. Chem. **265**, 3587–3590

CMC (= critical micelle concentration)

CNS (= central nervous system)

coated pit (*see* coated region)

coated region The region of the plasma membrane of a cell capable of endocytosis that contains receptors and is lined on the cytoplasmic side with the protein clathrin. When the receptors are occupied, these regions invaginate to form *coated pits* and then pinch off into *coated vesicles* in the cytoplasm. (*see also* receptosome)

coated vesicle (*see* coated region)

coatomer A complex of four proteins that coats Golgi-derived non-clathrin-coated vesicles, and that presumably supports their function in intracellular transport and budding.

• Waters, M.G., Serafini, T. and Rothman, J.E. (1991) Nature (London) **349**, 248–251

coat protein One of the proteins that encapsulate the nucleic acid core of a virus. (*see also* capsid)

cocktail A premixed reagent, i.e. a solution of phosphors for scintillation counting.

code blocker (antisense drug) In an attempt to develop specific therapies, free of side-effects, an antisense oligonucleotide or analogue that is designed to form a double-stranded complex with an mRNA produced by a virus or other pathogen, or during inflammation or oncogenesis. In the absence of the blocker, the targeted mRNA would otherwise direct synthesis of a protein that supports the infection, inflammation or oncogenesis. (*see also* antisense)

coding strand (*see* antisense)

codon The three-nucleotide sequence of an mRNA molecule that codes for one specific amino acid.

coenzyme A co-substrate in some enzymic reactions that is usually present in limited quantities *in vivo* and which requires regeneration in subsequent reactions, e.g. coenzyme A, NAD$^+$, FAD.

coenzyme I Obsolete name for NAD$^+$.

coenzyme II Obsolete name for NADP$^+$.

cognisable Descriptive of a ligand that is specific for a binding site, e.g. an inhibitor of an enzyme.

- Hakoshima, T., Itoh, T., Gohda, K., et al. (1991) FEBS Lett. **290**, 216–220

cohesive ended (= sticky ended)

coiled coil Several α-helices twisted together into a stout rope, e.g. in myosin, fibrin. (*see also* superhelix)

- Cohen, C. and Parry, D.A.D. (1994) Science **263**, 488–489

coincidence cloning (*see* difference cloning)

coincidence painting (chromosome painting) A method for production and utilization of fluorophore-labelled oligonucleotides to identify very localized regions of chromosomes. The method of *degenerate oligonucleotide primed PCR* (DOP-PCR) generates an assortment of characteristic fragments by use of a primer with arbitrary but fixed 3'- and 5'-ends, but randomized internally, so that it will find an appropriate number of hybridization sites on both strands of a source DNA to generate a convenient number of PCR fragments; in a selection step, the DOP-PCR products are annealed to carefully selected and characterized chromosome fragments fixed to a solid support, washed to remove non-hybridizing fragments, and the successfully annealed oligonucleotides eluted and labelled to serve as 'paint' that will colour and thus identify the chromosome region that selected them.

- Bailey, D.M.D., Carter, N.P., de Vos, D., et al. (1993) Nucleic Acids Res. **22**, 5117–5123; Wienberg, K. and Stanyon, R. (1995) Curr. Opin. Genet. Dev. **5**, 792–795

co-linearity The correspondence observed between the nucleotide sequence of a structural gene and the amino acid sequence of its protein product.

coliphage (*see* bacteriophage)

collected substrate inhibitor (= bi-product analogue)

colligative Descriptive of a property of a solution that depends upon the number but not the nature of solute molecules, e.g. osmotic pressure.

collisional limit (*see* spare receptors)

colloid A dispersion of a high-molecular-mass solute in a liquid phase. Typically, the colloidal solute is unable to traverse a semi-permeable membrane. The solute particles are usually of sufficient size to have an interior and a surface.

colony The progeny of a single micro-organism grown in culture on solid media and visible as a spot on a plate.

colony hybridization A technique for screening bacterial colonies for those that contain a desired polynucleotide sequence. A plate is exposed to a labelled oligonucleotide probe with a sequence complementary to part of the desired sequence which thus labels clones with that sequence. In a variant technique, *plaque hybridization*, phage vectors are screened for the desired polynucleotide sequence.

colostrum The secretion of the mammary gland immediately following parturition, that differs from milk in its content of immunoglobulins and other specialized proteins.

colour complementation assay (= colour PCR)

colour PCR A technique for screening a population for a specific point mutation that uses fluorophore-labelled PCR primers. Three primers are designed for amplification of a restriction fragment that contains the mutation site near one end: the primer for the end distal to the mutation site is unlabelled; two primers for the other end containing the normal and the mutant sequence are each labelled with a different and distinguish-able fluorophore, e.g. the normal with a red-fluorescing group and the mutant with a green-fluorescing group. When the restriction fragment is amplified in the presence of the primers, and the product separated from excess unincor-porated primers, the amplified products will fluoresce red if from a normal sub-ject, green if from a homozygous affected individual, and yellow if from a het-erozygous individual.

combinatorial Descriptive of procedures or processes that generate a diversity of molecules (e.g. nucleic acid or protein sequences; related chemical structures) that can be screened for a desired prop-erty. The collection of molecules is termed a *library*. (*see also* small mole-cule library)

combinatorial library (*see* repertoire cloning; synthetic peptide combinatorial library)

committed step The first irreversible enzymic reaction in a metabolic path-way, usually controlled by an allosteric enzyme.

comparative map An application of the comparative mapping of genes and other genetic markers in the genomes of sev-eral mammalian species to the localiza-tion to a specific chromosomal locus of a gene, in particular one thought to be related to a genetic disease (a candidate disease gene). Because mammalian evo-lution has resulted in relatively few gene relocations, and since those that have occurred seem to be random, knowledge of linkages of genes in mammalian species may be helpful in locating a gene in another mammal's genome.

• Eppig, J.T. and Nadeau, J.H. (1995) Curr. Opin. Genet. Dev. **5**, 709–716

compartment A functionally isolated reser-voir of a metabolite; the space occupied by a pool.

competitive binding assay A technique for assaying a substance by observing how effective it is, compared with standards, in displacing a fixed amount of the labelled (often radiolabelled) material from a binding site of great specificity and high affinity, usually an antibody or cell surface receptor. (*see also* ELISA; radioimmunoassay)

competitive inhibition A form of enzyme inhibition in which the inhibitor competes with the substrate for the enzyme's substrate-binding site. The result is an increase in the K_m value while leaving V_{max} unaltered. (*see also* inhibitor; non-competitive inhibition; uncompetitive inhibition)

competitive labelling (differential chemical modification) A method for characterization of the microenvironment of an amino acid residue of a protein, e.g. its accessibility to solvent or its pK_a; other residues of the same amino acid serve as internal controls. Rates of reaction or ratios of reaction compared with control residues are observed; very limited amounts of reagent and extents of reaction maximize the competition for the reagent between similar groups on the protein and ensure that only unmodified protein reacts with reagent.

competitive PCR A variation of PCR to make it a semi-quantitative assay for a polynucleotide. A new restriction site is introduced into the polynucleotide to be assayed so that, when an unknown quantity of the original and a known amount of the modified template are used together, both will be copied but their products will be distinguishable. The products are quantified on the basis of binding a radio-labelled hybridization probe and incorporation of labelled primer or fluorescence of bound ethidium bromide; the ratio of their products is equated with the ratio of the original templates.

• Gilliland, G., Perria, S. and Bunn, H.F. (1990) in PCR Protocols: A Guide to Methods and Applications (Innis, M.A., Gelfand, D.H., Sninsky, J.J. and White, T.J., eds.), pp. 60–69, Academic Press, San Diego

complement system A series of blood proteins that are part of the humoral immune response system which when activated causes a cascade of enzymic activations and reactions that result in chemotaxis, phagocytosis and lysis of foreign cells and bacteria.

complementarity-determining region (CDR) A polypeptide sequence of a variable domain of an immunoglobulin that is particularly responsible for its recognition by lymphocytes. These short sequences interrupt and loop out from the *framework regions* (FRs), which are relatively invariant and form the basic structural β-sheet scaffolding of the domains.

• Zanetti, M. (1992) Nature (London) **355**, 476–477

complementary In nucleic acid chemistry, descriptive of the relationship between two polynucleotides that can combine in an antiparallel double helix; the bases of each polynucleotide are in a hydrogen-bonded inter-strand pair with a complementary base, A to T (or U) and C to G. In protein chemistry, the matching of shape and/or charge of a protein to a ligand.

complementary base The purine that can form hydrogen bonds with a pyrimidine, and vice versa, in a double-stranded polynucleotide, e.g. G with C and A with T. (*see also* complementary)

complementary DNA (= cDNA)

complementation cloning A method for identification of a DNA fragment and selection of the defective cell line upon which it confers resistance to a selective pressure, e.g. cells that are hypersensitive to ionizing radiation. Fractions of normal genomic cDNA are screened for ability to confer resistance on cells into which they are introduced. Different lines of defective cells that represent discrete genetic defects may be associated with different cDNA fragments.

complex carbohydrate An oligomer or higher polymer of more than one kind of sugar moiety, or a glycoside formed with a non-sugar compound. (*see also* glycolipid; glycoprotein; glycoside; heteropolysaccharide; proteoglycan)

complexity In nucleic acid chemistry, a measure of the unique sequences in a DNA preparation; it increases with chain length and decreases with the extent to which the sequences are repetitive; evaluated as the size of a DNA fragment with no repetitive sequences that has the same $C_0 t_{1/2}$ value. In protein chemistry it is a quantitative measure of the randomness of a polypeptide sequence, C, which ranges from 0 for a homopolymer to 1.0 for a perfectly random sequence. For example, the complexity of collagen is 0.23 and that of globular proteins is about 0.90; $C = \frac{1}{3} \sum_{i=1}^{20} f_i \times \ln(f_i)$, where f_i is the mol fraction of amino acid i

among all the amino acids represented in the protein. (*see also* $C_0 t$ unit; simplified)

- Clarke, N.D. (1995) Curr. Opin. Biotechnol. **6**, 467–472

complex-type carbohydrate One type of glycoprotein moiety that is attached to the β-amide nitrogen of an asparagine residue (N-linked) and which may contain sialic acid residues at the non-reducing ends. (*see also* hybrid-type carbohydrate; high-mannose-type carbohydrate)

concatenate (catenane) A structure formed by two or more interlinked closed circular (cc)DNAs; formation occurs between the rapid replication of a ccDNA and the slow nicking of one of the ccDNAs to release it from the others.

concerted reaction A chemical or enzymic reaction in which several operations, e.g. attachment of an incoming group and departure of a leaving group, occur simultaneously and on opposite sides of the reaction centre.

condensation reaction The formation of a carbon–carbon, carbon–nitrogen or carbon–sulphur bond.

configuration An arrangement in space at an asymmetrical centre. (*see also* alpha-configuration; conformation)

conformation For a compound or macromolecule that has at least limited freedom of rotation about its chemical bonds, one alternative arrangement in space of its constituent atoms and groups. (*see also* configuration; tertiary structure)

conformational change The adjustment of a protein's tertiary structure in response to external factors (e.g. pH, temperature, solute concentration) or to binding of a ligand.

conformational hypothesis An explanation for oxidative phosphorylation that proposes that electron transport builds up a store of energy in the distorted shape of the mitochondrial inner membrane, the relaxation of which is coupled with phosphorylation of ADP. (*see also* chemical coupling hypothesis; chemiosmotic theory)

congenic (*see* transgenic)

connective tissue Support tissue that contains extracellular fibres such as collagen to give strength and protection to blood vessels, nerves, etc.; also bone and cartilage.

connexon (= hemi-channel)

Connolly surface (*see* solvent-accessible surface)

consensus sequence The minimal common sequence that appears in homologous polynucleotides or proteins, or limited regions of these, e.g. the sequence around the active-site region of serine proteinases.

conservation In molecular biology, the preservation through time of some bases in the polynucleotide sequence of an evolving gene or of some amino acids in the sequence of an evolving protein.

conservative mutation Such a mutation results in an amino acid substitution that preserves an essential chemical characteristic of the original, e.g. a leucine for an isoleucine, an aspartate for a glutamate, a lysine for an arginine.

constant region The part of the amino acid sequence of any one immunoglobulin that is the same for all immunoglobulins of its class. (*see also* hypervariable region; variable region)

constitutive enzyme An enzyme that is produced at a constant, non-inducible, rate. (*see also* inducible enzyme)

contact inhibition The cessation of division that occurs when cells in culture reach confluence and establish gap junctions with neighbouring cells.

contact number A measure of the location of an amino acid residue in a protein molecule; the number of α-carbons, excluding those of the adjacent residues in the linear sequence, within an 8 Å radius of the α-carbon of any given residue; this varies from 0 for residues on a protein's surface to about 15 for those in the interior. Because there is a higher proportion of internal to external residues in larger globular proteins, the values can be normalized by calculation of a *relative contact number*, i.e. the difference between the average contact number for all residues of a protein and that of any given residue, which varies between -5 and +5.

contact site A small region where the mitochondrial inner and outer membranes touch, and at which structures exist that are responsible for transport of proteins and adenosine nucleotides into and out of the mitochondrion.

- Moynagh, P.N. (1995) Essays Biochem. **30**, 1–14

contig map A characterization of a chromosome or large portion of one by *contiguous sequences*, which are zones of overlap of partially characterized segments; especially used as a strategy for co-ordination of efforts in many laboratories to sequence the human genome. (*see also* linkage map; physical map; restriction map; sequence-tagged site)

control (*see* regulation)

convertase A cellular proteinase that processes hormone precursors by recognition of the precursor sequence pairs of basic residues, which are the sites of cleavage.

• Marx, J. (1991) Science **252**, 779–780

converting enzyme (= convertase)

co-operativity (*see* negative co-operativity; positive co-operativity)

core glycosylation The attachment to a protein and modification of N-linked carbohydrate moieties that occurs in the endoplasmic reticulum (*see also* terminal glycosylation)

core protein One of the proteins to which are attached the carbohydrate chains (e.g. chondroitin sulphate, keratan sulphate) of a proteoglycan, and which with them comprise the subunits of the proteoglycan. (*see also* link protein)

Cori cycle The transport of the precursor and product of glycolysis between exercising muscle and the liver, i.e. lactic acid from muscle to liver, and glucose from liver to muscle.

Cori ester Obsolete name for α-glucose 1-phosphate.

Cornish-Bowden plot A graphical method for determination of the type of enzyme inhibition and the dissociation constant for an enzyme–inhibitor complex (K_i) or for an enzyme–inhibitor–substrate complex (K_i'). The effect on the enzymic rate (v) is determined at two or more substrate concentrations (S) and over a range of inhibitor concentrations (I). In a plot of S/v versus I, the data for each substrate concentration fall on straight lines that intersect at $I = -K_i$ and $S/v = K_m/V_{max}$ (uncompetitive inhibition), or intersect on the abscissa ($S/v = 0$) at $I = -K_i$ (noncompetitive inhibition). For competitive inhibition, the lines are parallel. (*see also* Dixon plot; Easson and Stedman plot; Hunter and Downs plot)

correlation spectroscopy (COSY) A two-dimensional nuclear magnetic resonance technique for detection of connectivity of the resonance of two centres through the covalent bonds that join them. (*see also* nuclear Overhauser and exchange spectroscopy)

• Kay, L.E. (1995) Curr. Opin. Struct. Biol. **5**, 674–681

correlation time In electron spin resonance and nuclear magnetic resonance spectroscopy, a time constant for the interaction of an observed species with its environment, often related to the rate of decay of its observed property; affected by spin relaxation, rotational motion and chemical exchange.

corrin ring A porphyrin-like ring system that lacks a methene bridge between two adjacent pyrrole rings; the basic structure of vitamin B_{12} and the coenzymes derived from it.

cortex The outer layer of a structure, e.g. adrenal cortex, cell cortex. (*see also* medulla)

corticoid An adrenal cortical steroid.

corticosteroid (= corticoid)

cosmid A plasmid used to introduce DNA sequences that are much larger than is suitable for other vectors.

cosmid walking A strategy for sequence analysis of a large polynucleotide; chromosome walking using a cosmid.

co-suppression The inhibition of gene expression by both sense and antisense RNAs.

• Jorgensen, R. (1992) Ag Biotechnol. News Inform. **4**, 265–273

COSY (= correlation spectroscopy)

co-translational Descriptive of a process of protein modification, e.g. glycosylation, that is concurrent with protein synthesis itself, i.e. that begins before protein synthesis is complete. (*see also* post-translational modification)

co-transport The concerted transport of metabolites or ions across a membrane. (*see also* antiport; symport)

Cotton effect A feature of some optical rotatory dispersion spectra that is usefully correlated with conformational and structural features of a molecule. A sharp peak in optical rotation followed by a deep trough is seen as the wavelength decreases; the wavelength between the peak and the trough where the rotation is zero is the point of maximal absorbance. A *negative Cotton effect* is the reverse, i.e. a fall in rotation followed by a steep rise as the wavelength decreases.

C_0t unit A compound unit used in measurement of the complexity of a DNA preparation. The rate of annealing of DNA is controlled by the second-order rate of nucleation, expressed as $C_0/C = 1 + k_2 C_0 t$, where C is the concentration of single-stranded DNA at time t, C_0 is the original concentration of single-stranded DNA (both expressed as the molarity of nucleotides), and k_2 is the second-order rate constant; often represented as a plot of C/C_0 against $\log(C_0 t)$. $C_0 t_{1/2}$ taken from such plots is the value of $C_0 t$ when $C/C_0 = 0.5$.

countercurrent chromatography A variant of partition chromatography that does not use a solid support to stabilize the stationary phase. A horizontal helical tube is filled with the stationary phase which has been equilibrated with the mobile phase; the mobile phase with solutes to be separated is introduced into one end of the helical tube and fractions are collected from the other. Equilibrium between the phases may be enhanced by slow rotation of the tube around its axis.

countercurrent distribution A method for separation of compounds based on differing partition ratios between immiscible liquids. In a linear series of many units (each consisting of an upper and a lower phase) after equilibration of the materials to be separated between two phases of one unit, the upper phase is repeatedly passed along for re-equilibration with the adjacent lower phase; this creates, after many such separations, binomial distributions of each constituent of the system among the upper and lower liquid phases of the units.

countercurrent elutriation (= centrifugal elutriation)

coupled-enzyme assay The determination of a substrate or enzyme activity by coupling of one enzymic reaction with another, more easily detectable, reaction. The product of the first reaction is the substrate for the second. These assays are often designed to generate or consume a reduced nicotinamide-containing nucleotide, which is measurable by its ultraviolet absorption.

coupled phosphorylation The synthesis of the phosphate anhydride bonds of ATP using energy derived from the electron transport chain.

coupling between conformational fluctuations (*see* entropy effect)

coupling factor A protein that permits the synthesis of ATP driven by the energy made available by mitochondrial electron transport.

covalent catalysis A stage in some enzymic reactions in which one moiety of a substrate is attached by a covalent bond to the enzyme.

covalent modification As applied to enzymes, the regulation of activity by modifications that may be reversible (e.g. phosphorylation or adenylation) or irreversible (e.g. limited proteolysis). (*see also* post-transcriptional modification; post-translational modification)

C_3 photosynthesis Carbon dioxide fixation by the reductive pentose phosphate pathway, i.e. the Calvin cycle. (*see also* C_3–C_4 photosynthesis; C_4 photosynthesis)

C_4 photosynthesis (C_4 cycle; Hatch–Slack pathway) A variant of C_3 photosynthesis in which CO_2 is first concentrated at the site of photosynthesis by carboxylation of phosphoenolpyruvate to oxaloacetate, which is then transported as such or as an interchangeable 4-carbon dicarboxylic acid to the site of photosynthesis. This is an advantage to the plant because the key enzyme, ribulose bisphosphate carboxylase, which in photosynthesis forms 2 mol of 3-phosphoglycerate, has a competing activity, O_2 photorespiration, which produces both 3-phosphoglycerate and phosphoglycolate. (*see also* C_3–C_4 photosynthesis)

• Rawsthorne, S. (1992) Essays Biochem. **27**, 135–146

Crabtree effect The decrease in cellular respiration caused by increased glucose concentrations. (*see also* Pasteur effect)

crenarchaeote (*see* eukaryote)

crenote (*see* eukaryote)

Crick strand A designation of one arbitrarily chosen strand of double-stranded DNA to distinguish it from the other, called the *Watson strand*.

crista A deep indentation of the mitochondrial inner membrane.

critical micelle concentration (CMC) The minimum concentration of a detergent at which it will form micelles and below which it is a true solution.

crossing-over (*see* chiasma)

cross-link In protein chemistry, a natural or synthetic covalent bond between protein side chains; it can be directly between them or mediated by a spacer group. In carbohydrate and nucleic acid chemistry, a synthetic bridging group.

crossover The physical exchange of homologous parts between a pair of individual chromatids.

crossover connection A short polypeptide sequence that connects strands of β-structure.

crossover element A DNA sequence that is a preferred site of crossover between two genes or two chromosomes.

crossover experiment A method to detect the control point in a metabolic pathway by observation of the effect on metabolite levels as new steady-state levels are established after sudden creation or relieving of an experimental deficiency in a controlling factor, e.g. the effect of oxygen on the redox state of reduced electron transport components, or the effect of oxygen on the levels of glycolytic intermediates of anaerobic tissue.

cross-reacting material A protein product resulting from mutation that has lost its function but is recognizable by its ability to react with antibodies raised against the normal protein. More broadly, material may cross-react because it bears an epitope in common with the antigen.

cross-regulation Control of a metabolic pathway by the product of a different but related pathway, e.g. activation of the reduction of ADP to dADP by dGTP.

cryptogene (*see* guide RNA)

C-terminal In a polypeptide sequence, that unique residue which is connected to the linear sequence by its amino group, leaving it with a free carboxy group. In practice, the carboxy group of a C-terminal residue may be modified, e.g. by amidation or, in the case of pyroglutamate, by internal lactamization. (*see also* N-terminal)

C-terminal (carboxyl-terminal) analysis A chemical method for determination of C-terminal residues of proteins by cleavage by hydrazinolysis, which generates amino acid hydrazides of all amino acids except the C-terminal one; also an enzymic method using limited cleavage by a carboxypeptidase that sequentially liberates amino acid residues from the C-terminal position, i.e. the C-terminal residue first, the penultimate residue second, etc.

CTL Cytotoxic T-cell. (*see* T-cell)

culture An inoculum of cells, especially a pure strain, intended for propagation in liquid or on solid media; also the act of propagation of the cells.

cut Cleavage of the phosphodiester bonds of both strands of double-stranded DNA to produce blunt-ended fragments.

***C* value** The calculated value of the total DNA content of a cell per haploid number of chromosomes.

cycle sequencing A method that recruits PCR technology to assist sequencing of small amounts of a single-stranded DNA template. A heat-stable polymerase, a single primer, deoxynucleotides and small amounts of one of the four labelled dideoxynucleotides are incubated with the template in each of four incubations; in each of a series of cycles of annealing, polymerization and denaturation, a spectrum of single-stranded DNA fragments is generated that can be electrophoretically separated to produce a ladder from which can be assigned nucleotide positions relative to the primer.

cyclic amplification and selection of targets (CASTing) Essentially identical to the *selected and amplified (protein) binding site oligonucleotide (SAAB)* and *target detection assay (TDA)* procedures; a procedure for identification of consensus sequences of DNA to which a protein, e.g. a transcription factor, may bind. A random polynucleotide sequence is synthesized flanked by two defined sequences that will serve as templates for PCR primers; the polynucleotides are exposed to the DNA-binding protein, any complex that is formed is separated from the unliganded polynucleotides (e.g. by gel shift assay, affinity chromatography, filter binding) and the polynucleotide of the complex is isolated and amplified by PCR; repeated recycling through the sequence of ligand formation, selection and amplification results in a preparation that is sufficiently pure to be cloned into bacteria for larger-scale production. A variant is *systematic evolution of ligands by exponential enrichment (SELEX)* for identification of RNA sequences, which begins with a mixture of polyribonucleotides and in each cycle produces DNA from the selected RNA–protein complex using reverse transcriptase, amplifies it by PCR, and then produces new RNA transcripts for the next round of selection.

- CASTing: Wright, W.E. and Funk, W.D. (1993) Trends Biochem. Sci. **18**, 77–80; SAAB: Blackwell, T.K. and Weintraub, H.W. (1990) Science **250**, 1104–1110; SELEX: Turek, C. and Gold, L. (1990) Science **249**, 505–510; TDA: Thiesen, H.-J. and Bach, C. (1990) Nucleic Acids Res. **18**, 3203–3209; Ouellette, M.M. and Wright, W.E. (1995) Curr. Opin. Biotechnol. **6**, 65–72

cyclic nucleotide An internal nucleoside phosphodiester; usually a 2′,3′-diester or a 3′,5′-diester.

cystatin A term applied both to the superfamily of cysteine proteinase inhibitors and to one subgroup; the other subgroups are the kininogens and stefins.

cysteine proteinase (thiol proteinase) A type of peptidase that has at its active site a cysteine residue. (*see also* aspartate proteinase; metalloproteinase; serine proteinase)

cysteine proteinase inhibitor (*see* alpha$_1$-cysteine proteinase inhibitor; kininogen)

cysteine switch (= Velcro mechanism)

cyto- A prefix signifying cell.

cytoflow An instrument for flow cytometry.

cytoplasm The part of a cell inside the plasma membrane and outside the nucleus, comprising membrane-bound organelles, cytoskeleton, ribosomes and cytosol.

cytoplasmic streaming Concerted internal movement in some large cells in which bands of cytoplasm move just under the cell surface.

cytoskeleton The network of relatively rigid structures within a cell that give it shape and provide a framework against which intracellular movement may take place; includes microtubules, intermediate filaments, F-actin filaments and associated proteins.

cytosol The soluble part of a cell's cytoplasm, i.e. that part that does not sediment during ultracentrifugation.

cytosolic receptor As distinguished from cell surface receptors, an intracellular receptor protein, especially one for a steroid hormone. (*see also* nuclear receptor)

cytotoxic T-cell (*see* T-cell)

D

D- A symbol that indicates the absolute configuration of some asymmetrical metabolites, especially amino acids and sugars, and relates them to the configuration of (*d*)-glyceraldehyde. The opposite configuration is designated L-. α-Amino acids are related to glyceraldehyde as follows: hydrogen to hydrogen, carboxy to aldehyde, amino to hydroxy and side chain to methyl group. Sugars are classified according to the asymmetrical centre most remote from the carbonyl group of the molecule, e.g. C-5 of glucose. Note that absolute configurations are indicated as D- and L-. *d*- (*dextro*) and *l*- (*laevo*) refer to the direction of rotation of plane polarized monochromatic light, usually at 589 nm.

Da (= dalton)

dAB (= single-domain antibody)

dalton (Da) The unit of measurement of molecular mass, in g/mol.

Danielli–Davson model A model for biological membranes; a variation of the Gortner and Grendel model in which globular proteins are tightly adsorbed to the polar groups of the exposed surfaces of the membrane. (*see also* fluid mosaic model; unit membrane Danielli–Davson model)

dark reaction (= Calvin cycle)

dark repair (*see* photo-activated repair)

ddF (= dideoxynucleotide fingerprinting)

ddNTP Dideoxynucleoside triphosphate. The letter N refers to any or all of the common bases.

DEAD box protein One of a group of proteins that have in common an Asp-Glu-Ala-Asp (DEAD in the one-letter code) or related sequence, and function in ATP-dependent processing of RNA.
- Wasserman, D.A. and Steitz, J.A. (1991) Nature (London) **349**, 463–464

deamination The abstraction of the elements of ammonia from a compound, e.g. from histidine by the histidine lyase reaction, or from AMP in the adenylate deaminase reaction.

death gene A gene whose expression is associated with apoptosis, e.g. a Ca^{2+}-activated endonuclease that cleaves exposed regions of chromatin to produce nucleosome-sized fragments. (*see also* oncogene)

decorated thin filament A thin filament of muscle, i.e. F-actin, to which are attached globular heads. The S1 fragments of myosin resemble in electron micrographs a string with attached arrowheads, all angled towards one end of the string.

degeneracy Redundancy of the genetic code, in that each amino acid is specified by more than one codon.

degenerate oligonucleotide primed PCR (DOP) (*see* coincidence painting)

dehydro- A prefix that signifies a product of the removal of two hydrogen atoms, e.g. dehydroepiandrosterone.

dehydrogenation The oxidation of a compound by removal of equal numbers of protons and electrons, usually two of each.

delayed-early stage (*see* immediate-early stage)

deletion map The linear arrangement of natural mutations along a chromosome or part of a chromosome. In the first phase many individuals are studied for the presence of cytogenetic markers; later phases use restriction fragment length polymorphisms to detail the chemical nature of the aberrations.

- Vollrath, D., Foote, S., Hilton, A., et al. (1992) Science **258**, 52–59

deletion mutation A mutation caused by the absence of one of more nucleotides in the DNA sequence.

delta restriction cloning A method for the use of a single primer to determine a DNA sequence longer than would normally be practical for a single sequencing attempt. After the first sequencing the bulk of the determined sequence is excised and the plasmid re-ligated, so that the primer is just upstream from the remaining undetermined sequence. The procedure may be repeated indefinitely until a large sequence is determined.

denaturation The destruction of the ordered folding of a protein or nucleic acid that is required for its normal function. Protein denaturation often involves a change from a specific globular or fibrous conformation to a random coil; nucleic acid denaturation often involves the dissociation of a duplex into single strands. (*see also* native structure)

denaturing gel electrophoresis (= sodium dodecyl sulphate/polyacrylamide-gel electrophoresis)

denaturing gradient gel electrophoresis (DGGE technique) A method for detection of single base substitutions in DNA fragments. Putative mutant and normal double-stranded DNA fragments are applied along one edge of an agarose gel slab that contains a denaturing agent (e.g. urea and formamide) in a gradient perpendicular to the direction of electrophoresis. At a low concentration of the denaturant the fragments are more mobile, and at high concentration they are unfolded and therefore less mobile. The electrophoretic band describes a sigmoid curve on the developed gel; the denaturant concentration represents the mid-point of the curve and is characteristic of the sequence. Introduction of a GC-rich sequence, which is very stable to denaturation and thus acts as a *GC clamp*, into the DNA allows detection of mutations in the more stable regions of the DNA.

- Rossiter, B.J.F. and Caskey, C.T. (1990) J. Biol. Chem. **265**, 12753–12756

dendrogram A phylogenetic tree.

de novo synthesis The biosynthesis of a compound or chemical group from dissimilar compounds or groups, e.g. the formation of methyl groups from formyl groups.

density-gradient centrifugation A technique to separate and/or characterize a macromolecule by high-speed centrifugation in a density gradient formed by the concentration gradient of a solute such as CsCl or sucrose. The macromolecule sediments until it reaches the zone of its own density. Also known as isopycnic centrifugation or zonal centrifugation.

deoxy- A prefix that signifies a product of the replacement of a hydroxy group by a hydrogen atom, e.g. deoxycorticosterone, deoxyribose.

deoxyribonucleic acid (*see* DNA)

depletion–insertion A technique used to define properties of a cellular, subcellular or tissue model, in which a variable or undesired constituent, e.g. membrane phospholipid, is removed (by a phospholipase) and then replaced by a substitute constituent of defined characteristics (e.g. phosphatidylserine).

depside An ester of two or more derivatives of orsellinic acid (1,2-dihydroxy-5-methylbenzoic acid). A *depsidone* is a depside in which the aromatic subunits are additionally coupled to each other by an ether linkage. (*see also* acetate rule; acetogenin)

depsidone (*see* depside)

depurination The cleavage of N-glycosidic bonds of DNA to form apurinic DNA.

desensitization The loss of responsiveness of an enzyme to allosteric regulation while retaining its catalytic activity.

desmolase An enzyme that oxidatively cleaves carbon–carbon bonds, e.g. the side-chain-cleaving enzyme that converts cholesterol into pregnenolone.

desmosome A type of junction that attaches one cell to its neighbour.

desorption (*see* mass spectrometry)

desoxo- A prefix that signifies a product of the replacement of an aldehyde or ketone carbonyl oxygen with two hydrogen atoms.

desoxy- (= deoxy-)

destruction box A sequence of about 10 amino acid residues that determines a protein's susceptibility to ubiquitinization and subsequent proteolysis.

• Hunt, T. (1991) Nature (London) **349**, 100–101

detergent (surfactant) An amphipathic compound able to stabilize suspensions of non-polar materials in aqueous solution.

detoxification The chemical modification by oxidation, methylation, glycosylation, etc. of a xenobiotic to render it innocuous.

dextran A branched-chain storage polysaccharide of microbial origin.

dextrin (*see* Schardinger dextrin)

***dextro*-rotation** (*see* optical rotation)

DGGE technique (= denaturing gradient gel electrophoresis)

diagonal electrophoresis A method for identification of a particular kind of peptide in a mixture by identical electrophoretic steps, the second at a 90° angle to the first, with a chemical modification introduced between the steps. An example is the identification of tyrosine-bearing peptides in a mixture of peptides by treatment of the products of paper electrophoresis with iodine vapour to iodinate the tyrosine residues before the second electrophoresis. Iodotyrosine peptides are then identified as those that deviate from the diagonal formed by all the other peptides when they are visualized by, for instance, the ninhydrin reaction.

dialysis A technique for the separation of macromolecules from smaller molecules by placing them within a semi-permeable membrane, such as Cellophane, separating them from a large volume of water. Only the low-molecular-mass diffusible molecules cross the membrane and pass into the larger volume; the macromolecules are confined to their original space. *Equilibrium dialysis* is the technique of quantification of binding capacity and affinity by dialysis of a macromolecule against various concentrations of a ligand and subsequent measurement of the final concentrations of bound and free ligand within the dialysis chamber and free ligand outside it.

diamagnetic Descriptive of a compound or chemical group (usually one that contains no unpaired electrons) that is not affected by a magnetic field. (*see also* paramagnetic)

diaphorase An enzyme that transfers electrons from NADH to a dye or to ferricyanide.

diastereoisomer One of two or more compounds that differ from each other at one or more asymmetrical centres, e.g. D-erythrose and D-threose. (*see also* enantiomer; epimer; stereoisomer)

diauxie Bacterial growth in two stages which occurs when the bacteria are grown on two carbon sources, e.g. glucose and xylose. The first phase corresponds to consumption of one compound, followed by a lag phase until enzymes for the assimilation and metabolism of the second compound are produced.

• Roseman, S. and Meadow, N.D. (1990) J. Biol. Chem. **265**, 2993–2996

Dickens–Warburg pathway (= pentose phosphate pathway)

Dictyostelium discoideum A cellular slime mould; the subject of much experimentation because it offers a mass of cells with synchronized cell cycles.

dideoxynucleotide A 2′,3′-dideoxynucleoside 5′-triphosphate; a deoxynucleotide analogue that lacks a hydroxy group at its 3′-carbon and functions as a chain-terminator during DNA synthesis. (*see also* Sanger method)

dideoxynucleotide fingerprinting (ddF) A method for the detection of mutation in a DNA fragment. The DNA serves as a template for replication using a mixture of four normal deoxynucleotides plus one dideoxyribonucleotide. Electrophoresis on a non-denaturing gel shows shifted mobility for the mutant template products compared with those of the normal DNA template. (*see also* singlestrand conformational polymorphism)

• Sarkar, G., Yoon, H. and Sommer, S.S. (1991) Genomics **13**, 441–443

dideoxynucleotide sequencing (= Sanger method)

diester (*see* phosphodiester)

difference cloning A group of techniques (e.g. subtractive DNA cloning) designed to isolate DNA sequences that are not shared between two DNA sources, e.g. cDNA clones prepared from different tissues, or under different conditions. Difference cloning is distinguished from *coincidence cloning*, comprising techniques (e.g. end ligation coincident sequence cloning) designed to isolate DNA sequences that are shared between two DNA sources.

difference Fourier method A technique of X-ray crystallography for solving the structure of a protein–ligand complex once the structure of the original protein has been deduced, by determination of the difference in distribution of densities and, eventually, the electron densities that result from altering a crystal of the protein by diffusing the ligand into it.

difference Patterson map An interim stage in the solution of a molecular structure from X-ray crystallographic data. Using only the intensities due to heavy atom reflections obtained by subtraction of the reflections of the protein from those of an isomorphous replacement, Fourier analysis calculates the vectors between heavy atoms. Their display is the difference Patterson map. This map may then be used to calculate the relative positions of the heavy atoms. (*see also* Fourier transformation; phase problem)

differential centrifugation The fractionation of subcellular components according to their sedimentation behaviour; separation into nuclei, mitochondria, lysosomes, microsomes (endoplasmic reticulum), ribosomes, cytosol, etc. by removal of sedimenting material after cycles of processing at progressively increasing centrifugal force.

differential chemical modification (= competitive labelling)

differential PCR display A method for identification of mRNAs produced under specific physiological conditions. Total cellular RNA is reverse-transcribed, and the resultant cDNAs are used as templates for PCR. The 3'-primer has a poly(T) sequence that directs it to the poly(A) tail of mRNA, and is 5'-ended with two bases that make this primer more selective. (The two bases may be varied to achieve different selections.) The 5'-primer is short and of arbitrary sequence, and is intended to allow, under controlled non-stringent reannealing conditions of temperature, concentration, time, etc., amplification of a manageable number of specific cDNAs (between 50 and 100 bands) that can be electrophoretically separated on a DNA sequencing gel. The ladder of cDNAs can be displayed side-by-side with the cDNAs derived from cells in a different physiological state. This allows the identification of unique cDNAs that can be extracted, amplified by PCR, sequenced and identified. In a variant version called *RNA fingerprinting arbitrary primer PCR (RAP-PCR)*, no poly(T) primer is used; one primer can serve for each of the two transcribed strands and the non-stringent reannealing conditions allow a fit to a manageable number of sites on the cDNA template.

• Liang, P. and Pardee, A. (1992) Science **257**, 967–971; McClelland, M., Mathieu-Daude, F. and Welsh, J. (1995) Trends Genet. **11**, 242–246

differential reassociation (*see* subtractive DNA cloning)

diffraction pattern The array of reflections obtained by crystallography. Each reflection indicates an intensity and, by its location, the angle with respect to the incident beam. Because a sharp pattern requires the wavelength of the incident beam to be of the same order of magnitude as the regular spacings, determination of inter-atomic distances is performed by X-ray crystallography, and the X-ray diffraction pattern is used to construct a three-dimensional model of the crystal. (*see also* phase problem)

diffusion coefficient A measure of a molecule's ability to travel through a solution, propelled by Brownian movement; expressed as RT/Nf (cm^2·s^{-1}), where R is the gas constant, T is the absolute temperature, N is Avogadro's number and f is the frictional coefficient.

diffusion limited Descriptive of a rate, e.g. of an enzymic reaction, that is determined only by the rate of diffusion of reactants towards the enzyme or of products away from it.

digestibility The characteristic of a protein that accounts for its nutritional quality according to how efficiently its digestion delivers its amino acids, especially the essential amino acids, to the individual.

digestion The degradative action of hydrolytic enzymes or of reagents such as acids.

dihydro- A prefix that signifies a product obtained by the addition of two hydrogen atoms to a precursor.

diketopiperazine A six-membered heterocyclic product of the condensation of two α-amino acids that contains amide bonds between the two α-amino and the two α-carboxy groups.

dimedone 5,5-Dimethyl-1,3-cylcohexanedione; a reagent that has been used to assist in the isolation of aldehydes and in their quantification (e.g. for [^{14}C]formaldehyde by its radioactivity) because it forms a water-insoluble, crystallizable adduct.

Dintzis experiment The method by which it was demonstrated that proteins are synthesized in the N- to C-terminal direction. A labelled amino acid was supplied to a reticulocyte preparation synthesizing haemoglobin and the completed globin chains were isolated after a brief incubation. The radioactive products were found to be concentrated in the C-terminus, indicating that this was the last part of the molecule to be formed.
- Dintzis, H.M. (1961) Proc. Natl. Acad. Sci. U.S.A. **47**, 247–261

dioxygenase An enzyme that reduces molecular oxygen by incorporating both atoms into its substrate, e.g. tryptophan dioxygenase. (*see also* mixed-function oxygenase)

diploid Descriptive of the number of chromosomes of a somatic cell, i.e. two of each chromatid. (*see also* haploid)

direct bilirubin (*see* van den Bergh reaction)

direct calorimetry Evaluation of the heat evolved by a human or experimental animal by measurement of the heat exchanged with the environment in specially constructed insulated chambers. (*see also* indirect calorimetry)

directed proton transfer (*see* entropy effect)

direct van den Bergh reaction (*see* van den Bergh reaction)

disc-gel electrophoresis A technique for electrophoresis of single samples in an open-ended tube. The polyacrylamide stationary phase is polymerized *in situ*. During electrophoresis the ends of the tube are immersed in upper and lower buffer chambers through which it is connected to the power supply. The highly focused gel travels through the tube as a disc. (*see also* slab gel)

disease gene mapping The localization of a specific gene to a locus on a chromosome by study of the morphology of chromosomes and by linkage studies.

displacement chromatography A column chromatographic method for separation and concentration of components of a solution. After adsorption of the sample to be separated on to the resin in a column, all bound molecules are displaced by a concentrated solution of a displacer, whose affinity for the stationary phase must be higher than that of any of the sample molecules. As the displacer sweeps the adsorbed molecules ahead of itself, each of these becomes a displacer for any molecule of lower affinity. Thus the original components are eluted in small volumes ahead of the displacer, in reverse order with regard to their affinities for the stationary phase.

dissociation constant (K_d) Given by $[A][B]/[AB]$, where $[AB] = [A] + [B]$; expressed in units of concentration. (*see also* association constant; pK_a)

distributive control The principle that the control of flux of metabolites through a pathway resides in several enzymic steps and not, as proposed by simpler models, in a single rate-controlling step.
• Srere, P. (1994) Trends Biochem. Sci. **19**, 519–520

disulphide bridge An inter- or intra-polypeptide cross-link formed by oxidation of the thiol groups of two cysteine residues to a single cystine residue.

diterpene (*see* terpene)

Dixon plot A graphical method for determination of the type of enzyme inhibition and the dissociation constant (K_i) for an enzyme–inhibitor complex. The effect on the enzymic rate (v) is determined at two or more substrate concentrations, and over a range of inhibitor concentrations (I). In a plot of $1/v$ against I, data for each substrate concentration fall on straight lines that intersect at $I = -K_i$ and $1/v = 1/V_{max}$ (competitive inhibition), or that intersect on the abscissa ($1/v = 0$) at $I = -K_i$ (non-competitive inhibition). For uncompetitive inhibition, the lines are parallel. (*see also* Cornish-Bowden plot; Easson and Stedman plot; Hunter and Downs plot)

D-loop A structure of heteroduplex polynucleotides in which one strand is longer than the other, forcing the internal extra nucleotides into a loop. (*see also* heteroduplex mapping)

DMC (= double minute chromosome)

DNA Deoxyribonucleic acid; a macromolecule formed of repeating deoxyribonucleotide units linked by phosphodiester bonds between the 5'-phosphate group of one nucleotide and the 3'-hydroxy group of the next. DNA appears in Nature in both double-stranded (the *Watson–Crick model*) and single-stranded forms, and functions as a repository of genetic information that is encoded in its *base sequence*. (*see also* A-DNA; B-DNA; Z-DNA)

DNA fingerprinting (DNA profiling) A method to generate a pattern of DNA restriction fragments that is unique to an individual, especially for forensic purposes. DNA from blood, semen or another tissue sample is isolated and cleaved with a restriction endonuclease; the products are separated by polyacrylamide-gel electrophoresis, blotted on to a nylon sheet, and fragments that contain a specific nucleotide sequence are detected by hybridization with an appropriate DNA probe to produce the unique pattern. (*see also* variable number tandem repeat)

DNA gyrase An enzyme that uses the energy of ATP hydrolysis to unwind double-stranded circular DNA to form a negatively supercoiled molecule.

DNA profiling (= DNA fingerprinting)

DNA repair The removal of damaged segments, e.g. pyrimidine dimers, from one strand of double-stranded DNA and its correct resynthesis.

DNase footprinting (= exonuclease footprinting)

DNA typing (= DNA fingerprinting)

dNTP Deoxynucleoside triphosphate; the letter N refers to any or all of the common bases.

docking In a computer graphics simulation, the bringing together of two molecular models to explore their interactions, e.g. the insertion of a substrate into the active site of an enzyme.

docking protein (*see* signal recognition particle)

dogma (*see* central dogma)

domain A region of a globular protein that has its own tertiary structure, and that is stable independently of the rest of the protein; often connected to other domains of the same protein by short sequences without secondary structure.

dominant (*see* allele)

Donnan effect (Gibbs–Donnan effect) The unequal distribution of a diffusible ion across a semi-permeable membrane when an impermeable electrolyte, such as a protein, is also present on one side; e.g. a solution of an anionic protein with Na^+ as a counter-ion, in contact through a membrane with a NaCl solution, will result in the transfer of Cl^- into the protein compartment and an equivalent amount of Na^+ out of it, such that the products of concentrations, $[Na^+] \times [Cl^-]$, in each compartment are equal.

donor (*see* acceptor)

donor (5′) splice site The site on pre-mRNA that corresponds to the 3′-end of an exon and the 5′-end of an intron. The first two bases of the intron at the donor splice site are GU in most cases, but AU in some instances.

• Mount, S.M. (1996) Science **271**, 1690–1692

DOP Degenerate oligonucleotide primed PCR. (*see* coincidence painting)

dot blotting A method to estimate the concentration of a polynucleotide or protein solution by spotting it on a sheet of nitrocellulose or nylon, hybridizing it with a radiolabelled complementary polynucleotide or antibody, visualizing it by autoradiography, then comparing its intensity with that of a similarly treated graded concentration series of the standard polynucleotide or protein.

double-displacement kinetics (= Ping Pong Bi Bi kinetics; *see* enzyme mechanism)

double helix (Watson–Crick model) The arrangement in space of two polynucleotide chains in which each chain is wrapped around the other to form two antiparallel spirals. Each strand presents to the other the bases, purine to pyrimidine, with which it can form inter-strand hydrogen bonds. (*see also* A-DNA; Z-DNA)

• Watson, J.D and Crick, F.H.C. (1953) Nature (London) **171**, 737–738

double minute chromosome (DMC) An autonomously replicating structure found, for example, in leukaemic cells.

double-reciprocal plot (= Lineweaver – Burk plot)

double sieve Descriptive of the mechanism for assuring correct charging of a tRNA by its amino acid activating enzyme. Only an amino acid that exactly fits the binding site will be ligated to the nucleic acid; those that are too large do not bind to the active site and those that are too small are hydrolysed. (*see also* proofreading)

double-strand-break repair model A conceptualization of the general genetic recombination of newly replicated double-stranded DNA during meiosis. A nuclease cleaves both strands of one duplex, the 3'-overhang invades the other duplex, displaces one strand and pairs with the complementary strand at a homologous site. Repair synthesis elongates the 3'-end that has annealed to the complementary strand of the intact duplex and extends the length of the invasion of the intact duplex. Similarly, the other 3'-end of the cut duplex is extended by repair synthesis along the displaced strand, both 3'-ends are religated to the 5'-ends of the cut duplex, and the entire complex is resolved by cutting and then religating an inner or an outer strand into non-crossover or crossover duplexes. (*see also* Holliday model; Meselson–Radding model)

• Shinohara, A. and Ogawa, T. (1995) Trends Biochem. Sci. **20**, 387–391; Weaver, D.T. (1995) Trends Genet. **11**, 388–392

double-stranded Descriptive of two complementary polynucleotide strands paired in a double helix; also (less commonly) of a protein multimer, such as F-actin.

doublet A subunit of the axoneme in which 10 microtubule protofilaments are grafted longitudinally on to a whole microtubule, which itself consists of 13 protofilaments. Also, in magnetic resonance, a peak that has been split into two by the interaction of a resonance with a nearby perturbing centre. (*see also* splitting)

double-well hydrogen bond (*see* hydrogen bond)

downfield Descriptive of a resonance position in a magnetic field lower than that at which a standard displays its signal.

down-regulation The decrease in the number of hormone receptors on a target cell that occurs after exposure to the hormone. Also, in bacterial nutrition, equivalent to down-shift.

down-shift The response by cells, especially micro-organisms, to an altered nutritional state, e.g. the decrease in ribosomes or enzyme production when levels of some nutrients are low.

downstream In a polynucleotide chain, towards the 3' end. (*see also* upstream)

DPN$^+$ Diphosphopyridine nucleotide; obsolete name for NAD$^+$.

DPNH Reduced diphosphopyridine nucleotide; obsolete name for NADH.

driver (*see* subtractive DNA cloning)

Drosophila melanogaster The fruit fly; the subject of classical genetic studies, because of its rapid generation time and easily observed characteristics.

dsDNA Double-stranded DNA. (*see also* double helix)

dsRNA Double-stranded RNA.

dual-recognition model (*see* altered-self hypothesis)

duplex A double-stranded polynucleotide.

dUTP system mutagenesis (= Kunkel mutagenesis)

dyad symmetry A property of double-stranded DNA whereby one strand has the same base sequence (5′-to-3′) as its complementary strand (5′-to-3′), which allows it to bind to a homodimeric or homotetrameric binding protein, e.g. the *lac* repressor. In practice, the term is applied to polynucleotide sequences that are more extensive, but have less perfect symmetry, than the palindromic sequences that are sites of restriction nuclease cleavage.

dye primer (*see* Sanger method)

dye terminator (*see* Sanger method)

E

Eadie–Hofstee plot A linearized display of kinetic data of dependence of enzyme activity on substrate concentration; rate is plotted against the ratio of rate to substrate concentration. (*see also* Lineweaver–Burk plot)

Easson and Stedman plot A graphical method for determination of enzyme–inhibitor dissociation constants for non-competitive inhibitors, especially those that bind very tightly. Where I is the total inhibitor concentration, E is the total enzyme concentration and α is the ratio of inhibited to non-inhibited rates under the same conditions, I/α is plotted on the ordinate (y-axis) against $1/(I-\alpha)$ on the abscissa (x-axis) to give a straight line that intersects the ordinate at E; its slope is K_d, the dissociation constant. (*see also* Cornish-Bowden plot; Dixon plot; Hunter and Downs plot)

EC$_{50}$ The concentration of an effector that evokes a response in 50% of the subjects, animals, cells, etc. in which it is tested.

eccrine secretion (= exocytosis)

E. coli (= *Escherichia coli*)

ectoderm One of the three primordial germ layers formed during early embryogenesis; a precursor of the central nervous system, sensory organs, adrenal medulla, etc. (*see also* endoderm; mesoderm)

ectoenzyme A plasma-membrane-associated enzyme that catalyses reactions in the extracellular space, e.g. acetylcholinesterase, 5′-nucleotidase.

editing (= proofreading)

Edman degradation A chemical technique to degrade and cleave amino acid residues sequentially from a protein beginning at the N-terminus, and to identify the residues as they are removed. Reaction of the N-terminal residue with phenylisothiocyanate cleaves it from the protein as the phenylthiohydantoin derivative, which may be isolated and identified. (*see also* thiocyanate degradation)

• Edman, P. (1950) Acta Chim. Scand. **4**, 283–293

effector A compound that modulates an allosteric enzyme; a system that produces an intracellular response to a hormone, e.g. adenylate cyclase.

EF hand The region in some homologous calcium-binding proteins where the ion is located. The E and F helices of calmodulin, troponin C, etc., resemble the extended index finger (E helix) and thumb (F helix) of the right hand. The ion is located in the pocket formed by the closest approach of the E and F helices to each other.

eicosanoid One of a class of compounds that includes the prostaglandins, thromboxanes and leukotrienes, all derived from arachidonic (eicosatetraenoic) acid.

Eisenberg plot A graphical representation of the hydrophobic moment $<\mu_M>$ against the mean hydrophobicity $<H>$ of an α-helix. α-Helices of globular proteins generally have low $<\mu_M>$ and low $<H>$ values; membrane surface-associated helices have high $<\mu_M>$ and high $<H>$ values; and α-helices imbedded in membranes have low $<\mu_M>$ and high $<H>$ values.

EL-CSC (= end ligation coincident sequence cloning)

electrical breakdown (= electropermeabilization)

electric field effect (*see* entropy effect)

electroblotting An electrophoretic technique for transfer of a protein or nucleic acid from a slab gel, or similar unstable two-dimensional matrix, to a sheet of stable material such as nitrocellulose or poly(vinylidene difluoride), for further operations or analysis. The electric field is perpendicular to the plane of the gel and the sheet. (*see also* electroelution)

electrochemical gradient A concentration and charge difference across a membrane, e.g. the pH difference and membrane potential developed during mitochondrial electron transport. (*see also* chemiosmotic theory)

electroelution A technique to remove a sample previously purified by electrophoresis on a solid support, by electrophoresing it into a buffer. (*see also* electroblotting)

electrofuge A chemical group that is displaced in an electrophilic substitution or elimination reaction. (*see also* nucleofuge)

electromicrofiltration A separation procedure in which an electric field is imposed upon an ultrafiltration membrane in order to prevent non-permeable material from accumulating on and clogging it.

electron density map In X-ray crystallography, a representation that resembles a geological survey map of planes drawn through a crystal. It shows contours of equal electron density surrounding the individual atoms: steeper around heavier atoms; more shallow around lighter atoms.

electronegativity The degree to which an atom or chemical group holds its electrons in competition with the atoms or groups to which it is bonded.

electron nuclear double resonance (ENDOR) A technique for detection of coupling between electrons and nuclei in order to gain information about the valance electron distribution around a paramagnetic nucleus; observation of an electron spin resonance transition while applying radio frequency energy to effect nuclear magnetic resonance transitions.

electron paramagnetic resonance (= electron spin resonance)

electronic film (= area detector)

electronic strain The distorted distribution of electrons in a substrate induced by the proximity of a strong dipole in the enzyme, e.g. Zn^{2+} in carboxypeptidase.

electron sink In discussions of enzyme mechanisms, a group that can pull electrons from a reactive centre and thus stabilize an electron-deficient intermediate or transition state, e.g. pyridoxal phosphate condensed with an α-amino acid.

electron spin resonance (ESR) Also known as *electron paramagnetic resonance (EPR)*; a magnetic resonance technique that detects and characterizes the environment of a moiety that contains an unpaired electron by the energy that is equivalent to a change in its spin state. (*see also* spin label)

• Well, J.A., Bolton, J.R. and Wertz, J.E. (1994) Electron Paramagnetic Resonance: Elementary Theory and Practical Applications, John Wiley & Sons, New York

electron-transfer potential A quantitative measure of the ability of a molecule to lose an electron in a redox reaction under standard conditions; expressed in volts (V). (*see also* redox potential)

electron transport The mediation of oxidation of one metabolite and the reduction of another by a series of carriers, cytochromes, iron-sulphur proteins, quinones, etc.

electron transport chain (respiratory chain) Usually, mitochondrial electron transport.

electron transport particle (= submitochondrial particle)

electro-osmosis The movement of water with its solute ions under the influence of an electric field, e.g. the dehydration of a non-submerged agarose gel as its counterions are electrophoresed out of it.

electropermeabilization A technique for the introduction of materials into an intact cell. Cells are subjected to a strong electric field which causes them to become temporarily permeable and thus take up molecules from their environment that would otherwise remain excluded from them. Also known as electrical breakdown, electropermeation, electroporation and high voltage electrical discharge.

• Schulz, I. (1990) Methods Enzymol. **192**, 280–300

electropermeation (= electropermeabilization)

electropherogram A graphical representation of an electrophoretic separation; e.g. for a slab gel, absorbance (ordinate) plotted as a function of distance from the origin (abscissa)

electrophile A compound or functional group that can attract an electron pair. (*see also* nucleophile)

electrophoresis A technique that separates charged compounds according to their mobilities in an electric field. In *free zone electrophoresis* the compounds are in a solution unsupported by any stabilizing matrix. In *gel electrophoresis* the solution is stabilized by a gel (e.g. agarose, polyacrylamide, starch). In *paper electrophoresis* the stabilizing matrix is buffer-soaked paper. In *isotachophoresis*, by contrast, charged solute molecules all travel at the same rate: a highly mobile leading ion (e.g. Cl^-) travels first and a relatively immobile ion (e.g. tricine) is the trailing ion; to avoid an electrostatic gap, other solute molecules travel between them, all at the same rate but ordered according to their relative mobilities. *Capillary eletrophoresis* uses a long, very-narrow-bore tube coated with the stationary phase; as the mixture passes along the length of the lumen, components are differentially retarded to effect a separation. (*see also* rocket electrophoresis)

electrophoretic mobility shift assay (EMSA) (= gel shift assay)

electroporation (= electropermeabilization)

electrospray mass spectrometry (ES-MS) (*see* mass spectrometry)

electrostatic stabilization (*see* entropy effect)

electrostriction The contraction of a solution volume in the presence of ionizing solutes, due to the more compact organization of water molecules around the charged groups.

ELISA Enzyme-linked immunosorbent assay; a technique that combines the specificity of an immunoglobulin with the detectability of an enzyme-generated chromophoric product to quantify a macromolecule. The enzyme is covalently attached to the immunoglobulin; when the latter is adsorbed to an antigen, its presence is revealed by the enzymic generation of a chromophore.

elongation cycle The reactions of a ribosome that add one amino acid residue to the C-terminus of a growing polypeptide chain and move the ribosome three nucleotides towards the 3'-end of the mRNA.

elongation factor An accessory protein that is required for transfer of amino acid residues to the polypeptide chain during translation.

elution In chromatography, the washing out of an adsorbed material from a solid support, especially by conditions that assist displacement, e.g. salt concentration, altered pH.

elutriation centrifugation (= centrifugal elutriation)

Embden–Myerhof pathway The sequence of enzymic reactions that convert glucose into pyruvic acid; also called *aerobic glycolysis*. (*see also* glycolysis)

emission spectrum In fluorescence spectroscopy, emission as a function of wavelength, upon excitation at a fixed, shorter wavelength. (*see also* excitation spectrum)

EMSA Electrophoretic mobility shift assay (= gel shift assay)

enantiomer The mirror image of an asymmetrical compound. (*see also* diastereoisomer; epimer; stereoisomer)

endergonic Descriptive of a chemical reaction that consumes free energy. (*see also* endothermic; exergonic)

end-group analysis A technique to determine the C-terminal or N-terminal residue of a protein. (*see also* Edman degradation; thiocyanate degradation)

end ligation coincident sequence cloning (EL-CSC) A technique for selection of DNA sequences common to two DNA populations. Of the two populations, the one of lowest complexity (or one made least complex by the creation of representations) is used in excess over the other population; it is cleaved by a restriction endonuclease and the fragments are ligated to linker oligonucleotides that will serve as primers, one of which is biotinylated; the other DNA population is cleaved by the same endonuclease. The two populations are mixed, denatured and annealed, and biotinylated fragments (i.e. including those from the second population that are also represented in the first) are isolated on a solid support, based on the presence of biotin. (The non-shared fragments from the second source are washed away.) The recovered fragments are then exposed to 'capture' oligonucleotides that are complementary to the linker oligonucleotides and will therefore be placed adjacent to the strands that originated in the second population; these strands are then ligated to the capture oligonucleotides. PCR using primers complementary to the capture oligonucleotides amplifies these fragments, which can be released from the primers by the endonuclease.

- Brookes, A.J., Slorach, E.M., Morrison, K.E., et al. (1994) Hum. Mol. Genet. **3**, 2011–2017

endo- A prefix that indicates something internal, e.g. an endonuclease is an enzyme that cleaves internal phosphodiester bonds of a polynucleotide. (*see also* exo-)

endocrine Descriptive of secretion into the bloodstream, e.g. insulin secretion. (*see also* autocrine; exocrine; paracrine)

endocytosis A process by which extracellular matter is taken up by a cell; the plasma membrane invaginates and encloses the material in newly formed vesicles. In *phagocytosis* the extracellular material is particulate; in *pinocytosis* it is non-particulate. (*see also* exocytosis)

endoderm One of the three primordial germ layers formed during early embryogenesis; a precursor of the gastrointestinal system, digestive glands, liver, lungs, etc. (*see also* ectoderm; mesoderm)

endo-form (*see* envelope conformation)

endogenous Arising within the organism or cell. (*see also* exogenous)

endopeptidase A peptidase that cleaves a protein at an internal peptide bond. (*see also* exopeptidase)

endoplasmic reticulum (ER) An extension of the nuclear envelope that forms sheets of membranes that are generally parallel to the nucleus of a cell. (*see also* rough endoplasmic reticulum)

ENDOR (= electron nuclear double resonance)

endorphin Endogenous morphine-like peptide; one of a class of peptides that are derived from pro-opiomelanocortin by limited proteolysis and have analgaesic properties. The group includes the 31-residue β-endorphin, and two pentapeptides, Leu-enkephalin and Met-enkephalin. (*see also* enkephalin)

endosome (= receptosome)

endosymbiotic hypothesis The proposal that some specialized organelles of contemporary organisms, e.g. chloroplasts and mitochondria, arose during evolution from the invasion of a eukaryotic cell by a bacterium.

endothermic Descriptive of a chemical reaction that consumes heat. (*see also* endergonic; exothermic)

endotoxin (*see* exotoxin)

end-point assay An enzyme-based assay that measures the amount of material by the quantity of a substrate consumed or product formed over the course of a reaction. (*see also* kinetic assay)

energy charge (= adenylate energy charge)

energy minimization (*see* molecular dynamics)

energy spectrum A profile of the energy of the emissions from a radioactive atom; the fraction of total emissions from the decay of a radioactive atom, e.g. the β-emissions of ^{14}C as a function of the energy of those emissions.

energy transfer (*see* molecular ruler)

enhancer A 50–150 bp sequence of DNA that increases the rate of transcription of coding sequences. It may be located at various distances and in either orientation upstream from, downstream from or within a structural gene. The site binds cellular transcription factors, including steroid–receptor complexes. (*see also* promoter)

enhancing screen A device to enhance the sensitivity of autoradiography; the surface that is to be visualized is sandwiched between a photographic film and the screen, which is a film impregnated with a phosphor that is excited by the sample's γ-rays and which can enhance the primary emission from the sample with its own secondary emission.

enkephalin A pentapeptide isolated from the brain which has opiate properties, e.g. Met-enkephalin (Tyr-Gly-Gly-Phe-Met), Leu-enkephalin (Tyr-Gly-Gly-Phe-Leu). (*see also* endorphin)

enol The tautomeric form of a -CH_2-CO- group; -CH=COH-. In nucleic acid chemistry, the alternative representations of the purine and pyrimidine rings as fully aromatic, with the substituent oxygens as phenolic hydroxy groups (enols), or as parts of carbonyl groups.

enrichment (*see* mole percent excess)

enterohepatic circulation The movement of a metabolite, e.g. a bile acid, from secretion by the gall bladder into the lumen of the intestine, its absorption, transport to the liver, and re-secretion in the bile.

enthalpy A measure of the internal energy of a system, comprising binding forces, pressure, etc., expressed as J/mol (or cal/mol). (*see also* second law of thermodynamics)

Entner–Doudoroff fermentation The metabolism of glucose, e.g. by *Pseudomonas fluorescens*, via 6-phosphogluconic acid and 2-oxo-3-deoxy-6-phosphogluconic acid to pyruvate and glyceraldehyde 3-phosphate.

entropy A measure of the disorder or randomness of a system, expressed in J/mol per degree K (or cal/mol per degree K). (*see also* second law of thermodynamics)

entropy effect The acceleration of a reaction that occurs when reactive groups are constrained in a productive orientation, either intermolecularly as on an enzyme surface, or intramolecularly as in a model compound. This phenomenon has been referred to variously as antichimaeric assistance, approximation, catalytic configuration, the Circe effect, coupling between conformational fluctuations, directed proton transfer, electric field effect, electrostatic stabilization, FARCE (freezing at the reactive centres of enzymes), gas-phase analogy, group transfer hydration, orbital perturbation theory, orbital steering, propinquity effect, proximity effect, rotamer distribution, stereoelectronic control; stereopopulation control, substrate anchoring, togetherness, torsional stress and vibrational activation.

• Page, M.I. and Jencks, W.P. (1987) Gazz. Chim. Ital. **117**, 455–460

envelope conformation A conformation of a five-membered ring, e.g. a furanose, in which four ring atoms lie in a plane and C-2 or C-3 (2-*endo* or 3-*endo*) is out of the plane. (*see also* twist conformation)

enzyme One of a class of biological catalysts that is composed principally of a globular protein of one or more polypeptide chains. In some cases enzymes include covalently bound or tightly associated metal ions, prosthetic groups or carbohydrates; they range in molecular mass from around 10 000 to several hundred thousand Da.

enzyme-activated inhibitor (= suicide inhibitor; *see* mechanism-based inhibitor)

enzyme-linked immunosorbent assay (= ELISA)

enzyme mechanism An elaborate system of nomenclature has been proposed to describe, from the perspective of kinetics, the variety of enzyme mechanisms. *Uni*, *Bi*, *Ter* and *Quad* refer to the number of substrates or products (e.g. Bi Bi refers to an enzyme that is bi-reactant in both directions, and uni bi to an enzyme that is uni-reactant in the forward direction and bi-reactant in the reverse direction). *Ordered* describes a mechanism that has an obligatory sequence of addition of reactants (and dissociation of products), *sequential* a mechanism in which all reactants are added before any product is released, and *random* a mechanism where there is no obligatory sequence. *Ping Pong* describes cases in which one or more products are released before all of the substrates have added to the enzyme. *Iso* refers to situations in which the enzyme undergoes isomerization to another stable form during the catalytic cycle.

- Cleland, W.W. (1963) Biochim. Biophys. Acta **67**, 104–137

enzyme–substrate complex (Michaelis complex) The association of a substrate with an enzyme that is an obligatory intermediate in conversion of the substrate into the product of the enzymic reaction.

eocyte An extremely thermophilic sulphur-dependent archaebacterium. (*see also* eukaryote)

eosinophil A polymorphonuclear leucocyte containing granules that react with eosin, a histological stain for acidic substances.

epigenetic Descriptive of control phenomena superimposed upon DNA-sequence-based phenomena, such as genetic imprinting, tissue-specific development, etc.

epigenetic state The condition of a cell that has been programmed during early embryogenesis, e.g. into ectoderm, endoderm or mesoderm, to a developmental fate that will be expressed many generations later.

epimer A compound that differs from another by its configuration at only one asymmetrical centre. (*see also* diastereoisomer; enantiomer; stereoisomer)

episome A polynucleotide that can be transferred from one bacterium to another, either as a discrete structure or as a sequence incorporated into the host genome. (*see also* plasmid)

epithelium A sheet-like tissue that lines body vessels, cavities and surfaces, and can differentiate into secretory glands.

epitope The antigenic determinant; i.e. that part of an antigen to which an antibody is directed; it defines the specificity of the antigen and consequently binds to the receptor on a T-cell or to the antibody. (*see also* paratope)

epitope library (= random peptide library)

EPR Electron paramagnetic resonance. (= electron spin resonance)

equatorial (*see* axial)

equilibrium dialysis (*see* dialysis)

equilibrium labelling In metabolic studies, the exposure of a biological system (e.g. a tissue slice, a cell culture, mitochondria) to a radiolabelled metabolite over a relatively long time so as to establish a constant steady-state specific radioactivity in downstream metabolites, in order to observe the metabolic fate of individual atoms of the original metabolite. (*see also* pulse labelling)

equilibrium sedimentation ultracentrifugation A technique for evaluation of the molecular mass of a polymer by determination of the extent of sedimentation at which the force exerted upon a macromolecule in a centrifuge is balanced by its diffusion due to Brownian motion. (*see also* sedimentation-velocity ultracentrifugation)

ER (= endoplasmic reticulum)

ergotypic (*see* anti-ergotypic)

ERK Extracellular-signal-regulated protein kinase (= MAP kinase)

erythrocyte An enucleated, mature, red blood cell.

Escherichia coli (E. coli) An enteric bacterium that inhabits the human intestine; much used for experimentation.

ESI Electrospray ionization mass spectrometry. (= ES-MS; *see* mass spectrometry)

ES-MS Electrospray mass spectrometry. (*see* mass spectrometry)

ESR (= electron spin resonance)

essential amino acid An amino acid that is not synthesized by an organism at an adequate rate (or at all) from other amino acids or metabolites; therefore one that is a dietary requirement.

essential fatty acid A fatty acid that is a dietary necessity, e.g. the polyunsaturated fatty acids linoleic acid and linolenic acid.

EST (= expressed sequence tag)

estrogen (= oestrogen)

Eubacteria (*see* eukaryote)

Eucarya (*see* eukaryote)

euchromatin In cytology, the lightly staining regions of chromosomes that contain less-condensed chromatin and are the regions that are transcribed as RNA. (*see also* heterochromatin)

euglobulin (*see* globulin)

eukaryote Common name of a member of the *Eucarya*, one of the three domains of living organisms that are classified according to rRNA sequence homologies; they are further characterized by cells containing nuclei and membranes composed primarily of diacylglycerol derivatives. The other two domains are the *Eubacteria* (the bacteria), which have no nuclei (i.e. prokaryotes) and also have diacylglycerol derivative membrane components, and the *Archaea* (archaebacteria), which are also prokaryotes; they have a membrane composed of isoprenyl glycerol diether and tetraether lipids. Kingdoms within the *Eucarya* include plants, animals and fungi; kingdoms of the *Archae* are euryarchaeotes (or euryotes), which include methanogenic, halophilic, sulphur-reducing and some thermophilic organisms, and the crenarchaeotes (or crenotes), which include other sulphur-dependent organisms, thermoacidophiles and extreme thermophiles (eocytes).

• Woese, C.R., Kandler, O. and Wheelis,

M.L. (1990) Proc. Natl. Acad. Sci. U.S.A. **87**, 4576–4579

euryarchaeote (*see* eukaryote)

euryote (*see* eukaryote)

evolutionary tree (= phylogenetic tree)

EXAFS (= extended X-ray absorption fine structure)

exaptation (*see* junk DNA)

exchange protein (= phospholipid-transfer protein)

exchange reaction A partial enzymic reaction, especially when only one of two substrates is present, in which chemical groups or single atoms of the substrate equilibrate with the medium, with a cofactor or with one of the products, e.g. the acetate-dependent incorporation of $[^{32}P]P_i$ into ATP, catalysed by acetyl-CoA synthetase, in the absence of coenzyme A.

excimer In fluorescence spectroscopy, a dimer of a fluorophore that is stable only in the excited state, e.g. of aromatics such as pyrene or 2,5-diphenyloxazole (PPO, used for scintillation counting).

excision The removal of a section of double-stranded DNA that is faulty due to mutation or incorrect replication. The process includes *incision*, i.e. cleavage of the strand by an endonuclease and the action of a 5'-nucleotidase, followed by filling in by a polymerase and ligation. The *E. coli* DNA polymerase I has all these activities. (*see also* proofreading)

excision repair The removal of a segment of a DNA duplex that contains a thymine or other dimer, formed by ultraviolet-light-induced damage, and resynthesis and ligation of the excised sequence.

• Seeberg, E., Eide, L. and Bjoras, M.

(1995) Trends Biochem. Sci. **20**, 391–397

excitable membrane A membrane containing a receptor which transduces an external signal, e.g. a membrane containing acetylcholine receptors or rhodopsin.

excitation spectrum In fluorescence spectroscopy, emission at a longer, fixed, wavelength as a function of the wavelength of the exciting light. (*see also* emission spectrum)

exergonic Descriptive of a chemical reaction that generates free energy. (*see also* endergonic; exothermic)

exinuclease An endonuclease that functions to excise damaged bases, e.g. thymidine dimers.

exo- A prefix that indicates something external, e.g. an exopeptidase is an enzyme that cleaves N- or C-terminal residues from a polypeptide. (*see also* endo-)

exocrine Descriptive of secretion into ducts that open on to body surfaces rather than into the bloodstream, e.g. salivary and gastric secretions. (*see also* autocrine; endocrine; paracrine)

exocytosis A secretory mechanism whereby secretory vesicles fuse with the cell membrane and expel their contents to the exterior. (*see also* apocrine; endocytosis; holocrine)

• Morgan, A. (1995) Essays Biochem. **30**, 77–95

exogenous Arising from a source outside the organism or cell. (*see also* endogenous)

exon From *expressed region*, i.e. a region of a eukaryotic gene that encodes a sequence of amino acids; as opposed to an *intron*, i.e. an intervening region or sequence. Introns are excised before translation of the resultant RNA transcript.

exon skipping The natural editing of a defective RNA transcript that excises the exon containing the defect (e.g. a premature stop codon) and generates an mRNA that encodes a modified protein; the cause of some instances of reappearance of a modified protein product after gene knock-out.

exon trapping A method for recovery of exons from random segments of cloned genomic DNA that depends upon passage through a retroviral life cycle; nonviral genomic DNA is spliced and may then be excised as cDNA from the viral DNA using appropriate exonucleases.

- Duyk, G.M., Kim, S., Myers, R.M. and Cox, D.R. (1990) Proc. Natl. Acad. Sci. U.S.A. **87**, 8995–8999

exonuclease footprinting (DNase footprinting) A method for identification of a protein-binding region in a double-stranded DNA fragment. One 5′-end of the DNA fragment is labelled with ^{32}P and the DNA-fragment–protein complex is treated with a 3′-exonuclease to digest from the 3′-end until it meets the region protected by the DNA-binding protein. The labelled strand is characterized subsequently by its size to indicate the distance of the site of the binding protein from the 5′-end of the fragment. The technique is suitable for localization of a binding site to a specific region of a large DNA fragment, whereas *footprinting* gives the exact sequence of the binding site of a smaller fragment. (*see also* footprinting)

- Revzin, A. (ed.) (1993) Footprinting of Nucleic Acid–Protein Complexes, Academic Press, San Diego

exopeptidase A peptidase that cleaves a protein sequentially, starting at the N-terminal peptide bond (an *aminopeptidase*) or at the C-terminal peptide bond (a *carboxypeptidase*).

exoskeleton The chitin-containing rigid external shell of insects and crustaceans.

exothermic Descriptive of a chemical reaction that generates heat. (*see also* endothermic; exergonic)

exotoxin A disease-causing agent that is produced and secreted by a pathogen; contrasted with an *endotoxin*, which is an intrinsic component of the pathogen, e.g. the lipopolysaccharide of Gram-negative bacteria.

- Raetz, C.H.R. (1990) Annu. Rev. Biochem. **59**, 129–170

explant A small sample of living tissue, often consisting of different tissue types, that can be sustained in a relatively differentiated state under cell culture conditions. (*see also* cell culture)

expressed sequence tag (EST) A partial coding sequence isolated at random from a cDNA library; like a sequence-tagged site for mapping total genomic DNA, used for identification and mapping of coding sequences, for discovery of new genes and (by reference to sequence data banks) for discovery of identities with other genes.

- Venter, C. (1993) Nature Genet. **4**, 373–380

expression The production of a gene product, i.e. protein or RNA, from a gene; the manifestation of a genotype as a phenotype.

expression cloning As part of a repetitive process for screening a library, the sampling of cells according to the presence of the product of the gene of interest, e.g. the identification of the glutamate receptor in a clone of cells as evidence that the receptor gene is contained within the cells.

expression vector A plasmid or bacteriophage, used as a vehicle for transfer of genetic information to a host cell, in which the inserted gene is expressed as a protein, e.g. the λgt11 vector.

extein (*see* protein splicing)

extended X-ray absorption fine structure (EXAFS) A technique for investigation of the immediate environment of metal atoms in metalloprotein crystals or solutions, e.g. Fe–S bond distances in rubredoxin; the X-ray energy is varied and the fine structure of the absorption spectrum is recorded indirectly as fluorescent radiation.

• Diakun, G.P. (1990) Nature (London) **344**, 83–84

extinction coefficient (= molar absorption coefficient)

extremophile An organism that can live in an extreme environment, e.g. *barophiles* (or *piezophiles*) that live under high pressure, *psychrophiles* that live at low temperatures, and *thermophiles* and *hyperthermophiles* that live at high temperatures; contrasted with a *mesophile*, which is an organism that lives under moderate conditions.

extrinsic factor Vitamin B_{12}. (*see also* intrinsic factor)

extrinsic pathway The blood clotting cascade that is initiated by a tissue factor. (*see also* intrinsic pathway)

extrinsic protein (= peripheral protein)

ex *vivo* 'Removed from life', e.g. an explant, a tissue in culture.

F

FAB (= fast-atom bombardment)

Fab fragment A product of papain hydrolysis of an immunoglobulin G consisting of the variable region and some of the constant region of a heavy chain and an entire light chain, and containing a single antigen-binding site. (*see also* Fc fragment)

facilitated diffusion A transport mechanism that moves compounds or ions down a concentration gradient, and requires no energy; also known as accelerated diffusion or mediated transport. (*see also* active transport; passive diffusion)

FACS (= fluorescence-activated cell sorting)

factitious protein A product of genetic engineering; a protein designed for a specific purpose or for its expected properties.

FARCE Freezing at the reactive centres of enzymes. (*see* entropy effect)

fast-atom bombardment (FAB) mapping A technique for comparison of recombinant with natural proteins or with expected structures. A protein is cleaved by a specific enzymic (e.g. trypsin) or chemical (e.g. CNBr) method and the product is characterized by FAB mass spectrometry. (*see also* fast-atom bombardment mass spectrometry)

fast-atom bombardment mass spectrometry (FAB-MS) A technique for the analysis of protein sequence and structure. A high energy beam of atoms or ions (Cs or Xe) vapourizes, fragments and ionizes a protein solution, the mass spectrum of which gives details of the peptide sequence and post-translational modifications such as N-terminal acylation, glycosylation, phosphorylation and disulphide bridging. *Flow FAB* is a variant that analyses the continuous liquid stream from a high pressure liquid chromatography or capillary electrophoresis separation. (*see also* fast-atom bombardment mapping)

fast-twitch muscle (= white muscle)

fat Usually neutral fat; triacylglycerols.

fatty acid A carboxy group on an alkyl chain that is usually unbranched; may be a short-chain (alkyl group contains less than 7 carbons), medium-chain or long-chain (alkyl group contains more than about 11 carbons) fatty acid.

Fc fragment A product of papain hydrolysis of an immunoglobulin G consisting of parts of the constant regions of two heavy chains held together by a disulphide bridge, but excluding antigen-binding regions. (*see also* Fab fragment)

FDAT (= fluorescence-based DNA analysis technology)

FD-MS Field desorption mass spectrometry. (*see* mass spectrometry)

FDR (= functional divergence ratio)

feedback inhibition A form of metabolic control in which the end product of a pathway inhibits the enzyme, usually an allosteric enzyme, that catalyses the earliest irreversible reaction that is unique to the pathway. Also, the control a hormone exerts on the sequence of events that results in its synthesis and release, e.g. the action of thyroxine on secretion of thyrotropin (thryroid-stimulating hormone).

feedback regulation Control of a metabolic pathway by a metabolite of the pathway that acts in the direction opposite to metabolic flux, i.e. upstream or 'earlier' in the pathway. (*see also* feed-forward regulation)

feed-forward regulation Control of a metabolic pathway by a metabolite of the pathway that acts in the same direction as the metabolic flux, i.e. downstream or 'later' in the pathway, e.g. the activation of pyruvate kinase by fructose 1,6-bisphosphate. (*see also* feedback regulation)

Fehling's solution (*see* reducing sugar)

Fenton reaction The generation of a reactive, and potentially damaging, hydroxyl radical: $H_2O_2 + Fe^{2+} \rightarrow OH^\bullet + Fe^{3+}OH^-$.

Ferguson plot A graphical representation of electrophoretic mobility as a function of gel concentration, according to the relationship: $\log M_0/\log M = K_r C$, where M_0 is the free mobility, M is the mobility in a gel of concentration C, and K_r is a retardation constant related to molecular size.

fermentation The anaerobic degradation, usually by a micro-organism, of a sugar (or other source of energy) and biosynthetic intermediates, during which secondary metabolites may be produced.

ferric chloride test The chelation of Fe^{3+} by an α-oxo acid to give a coloured product, as a test for metabolic products of amino acids, e.g. with α-oxophenylpyruvate as an indirect test for excess levels of phenylalanine in phenylketonuria.

FFF (= field-flow fractionation)

FFID Fission-fragment-induced desorption. (= plasma desorption mass spectrometry; *see* mass spectrometry)

fibroblast A relatively undifferentiated collagen-secreting cell of connective tissue.

fidelity The property of an amino-acid-activating enzyme or a polymerase to correctly charge a tRNA or to correctly place a residue in a growing polypeptide or polynucleotide.

field desorption mass spectrometry (*see* mass spectrometry)

field-flow (field-force) fractionation (FFF) A group of techniques for resolution of mixtures of macromolecules or colloidal particles. An external driving force (e.g. gravitational, electrical, magnetic) is applied perpendicular to the direction of continuous flow of the mixture through an elongated chamber, and the deflection of each suspended species in the field directs it to a unique area on the accumulation surface of the chamber.

field-force fractionation (= field-flow fractionation)

field ionization mass spectrometry (*see* mass spectrometry)

figure-eight form (*see* Holliday model)

FI-MS Field ionization mass spectrometry. (*see* mass spectrometry)

fingerprint The unique pattern of ninhydrin-reactive spots found upon separation of a tryptic digest of a protein by paper electrophoresis in one dimension, followed by chromatography in a second dimension at a 90° angle. Also, any unique pattern that is generated by separation of fragments resulting from limited proteolysis of a protein or hydrolysis of a polynucleotide at specific cleavage sites.

finite decomposition A type of mathematical modelling of steady-state metabolic rates according to the incremental contributions of all controlling factors. (*see also* sensitivity)

first-order kinetics In enzymology, the

direct dependence of rate on the substrate concentration; seen when this concentration is significantly below the K_m. (*see also* zero-order kinetics)

Fischer convention A representation of the absolute configuration of an asymmetrical carbon atom, especially the carbons of a sugar, in which the substituents further from the observer are shown above and below and the substituents closer to the observer are shown to the right and left. (*see also* Haworth projection)

FISH (= fluorescence *in situ* hybridization)

fission-fragment-induced desorption (= plasma desorption mass spectrometry; *see* mass spectrometry)

flagellum A long extension from the surface of a cell, often used for motility. Bacterial flagella are composed of strands of protein; eukaryotic flagella have an axoneme core.

flame photometry An analytical technique to quantify small amounts of some elements, especially alkali metals, in solution; the solution is vapourized in a flame and the intensity of the characteristic spectral line emission produced by excitation of the metal atoms is detected by a photomultiplier. (*see also* atomic absorption analysis)

flanking sequence A DNA sequence located adjacent to a gene, either upstream from its 5'-end or downstream from its 3'-end.

flip-flop (transverse diffusion) The slow diffusion of membrane lipids from one leaflet of a lipid bilayer to the other.

flip-flop circuit The sequence of enzymic events that permits the periodic expression of either of two genes, but never both together, e.g. control of the expression of two of the flagellar proteins of *Salmonella.*

flippase A putative enzyme that transports membrane elements, e.g. the lipopolysaccharide of Gram-negative bacteria, from the inner to the outer leaflet of the lipid bilayer cell membrane.
• Raetz, C.H.R. (1990) Annu. Rev. Biochem. **59**, 129–170

flipping (*see* base flipping)

florigen A plant hormone that induces development of a flower.

flow cytometry The fluorescence-activated cell sorting technique applied to analysis and/or separation of cells by their physical and biological characteristics.

flow FAB Flow fast-atom bombardment. (*see* fast-atom bombardment mass spectrometry)

fluctuating embryo (*see* framework model)

fluid mosaic model (Singer–Nicolson model) A conceptualization of a plasma membrane as a lipid bilayer in which embedded proteins are mobile in two dimensions. (*see also* Danielli–Davson model; Gortner and Grendel model; unit membrane Danielli–Davson model)
• Singer, S.J. and Nicholson, G.L. (1972) Science **175**, 723–731

fluor (= fluorophore)

fluorescence The property of a compound or moiety of absorbing ultraviolet or visible light and re-emitting it nearly instantaneously at a longer wavelength; also the fluorescent emission itself.

fluorescence-activated cell sorting (FACS) A technique for separation of cells according to their fluorescence after attachment of a fluorophore to specific cells, e.g. with a fluorescent antibody. Droplets that contain no more than one cell are passed by a device that imposes an electrical charge on fluorescent droplets and deflects them into their own receptacle. (*see also* cytoflow; flow cytometry)

fluorescence-based DNA analysis technology (FDAT) Automated DNA sequencing using single-lane sequencing; adaptable to automated data collection.

fluorescence *in situ* hybridization (FISH) A method for visualization of a genetic marker on a chromosome by use of a fluorescently labelled polynucleotide probe that hybridizes to and indicates the locus of a gene on a chromosome during mitosis (metaphase). (*see also* interphase nucleus mapping)

fluorescence lifetime The average time that a population of fluorophores spends in the excited state before collapse to the ground state. The lifetime is equal to the reciprocal of the first-order rate constant for the decay of fluorescence; this constant is increased, and the lifetime decreased, by non-radiative interaction of the excited state with its environment.

fluorescence photobleaching recovery A technique to measure the mobility of membrane lipids by destruction of fluorophore-tagged molecules in a very small area of a membrane and measurement of the rate at which new molecules migrate into it from the non-affected margins.

fluorescence recovery after photobleaching (= fluorescence photobleaching recovery)

fluorescence resonance energy transfer (= resonance energy transfer)

fluorography A variant of autoradiography with improved sensitivity for detection of radioactivity of bands in slab gels. The weak β emission from ^{3}H, ^{14}C or ^{35}S excites a scintillant such as salicylate that in turn exposes the photographic film.

fluorophore (fluor) A chemical group responsible for the fluorescence of a compound or macromolecule.

flush ended (= blunt ended)

flux-generating step An enzymic reaction that limits the rate of subsequent steps in its metabolic pathway. In a steady state the flux of metabolites, expressed as μmol/min, is the same for all steps. The step is usually characterized by thermodynamic irreversibility, and the enzyme has a K_m that is low compared with levels of its substrate *in vivo*, thus insulating itself and the pathway from variations in substrate levels.

folding unit A nucleation site of protein naturation or renaturation.

Folin–Ciocalteu reagent A reagent that uses phosphomolybdotungstate for the colorimetric detection of proteins. (*see also* Lowry protein assay)

footprinting A group of techniques for identification of protein-binding sites on DNA. In *dimethylsulphate (DMS) footprinting*, for example, an end-labelled DNA fragment is partially methylated at guanidine residues with DMS, then exposed to the DNA-binding protein. The material is electrophoresed on a non-denaturing polyacrylamide gel to separate fragments that are methylated in the protein-binding sequence (and will no longer bind) from those that are methylated only elsewhere and will be less mobile due to the bound protein. The two gel bands are then eluted and cleaved with piperidine at the methylated sites. Comparison of the cleavage ladders of the two preparations will show unique bands due to methylation and cleavage at residues in the non-binding material that were essential for binding. In *protection footprinting* a DNA fragment with and without a binding protein is subjected to cleavage in non-binding areas, e.g. by partial methylation and cleavage by piperidine. The material is separated on a sequencing gel and the ladder of the protein-bound preparation will show a gap (the footprint) in its sequence where the protein has bound, compared with the ladder from 'naked' DNA. Protection footprinting may be applied to intact cells and compared with naked DNA *in vitro* as the control. In that case, ligand-mediated PCR is used to label and amplify desired areas of the DNA. Other methods of cleavage are by DNase I and by hydroxyl radicals (OH$^\bullet$). (*see also* genetic footprinting)

- Lakin, N.D. (1993) in Transcription Factors: A Practical Approach (Latchman, D.S., ed.), pp. 27–47, Oxford University Press, Oxford

forced copy-choice model One of two modes of retroviral recombination. In making a DNA transcript of the two genomic RNA strands of a retrovirus, when the polymerase encounters a break on one template strand it continues on the other strand. More generally, the term describes recombination that occurs during minus strand synthesis. The other mode of retroviral recombination is the *strand displacement–assimilation model*, or recombination during plus strand DNA synthesis: when synthesis occurs on an intact RNA template, one fragment of DNA produced during the semi-discontinuous synthesis is displaced by the DNA strand being continuously synthesized on the other RNA template. Both modes of recombination have been detected.

- Hu, W.-S. and Temin, H.M. (1990) Science **250**, 1227–1233

Förster transfer (= resonance energy transfer)

Fourier transformation An analysis based on Fourier's theorem that any periodic function may be reduced to a series of sine and cosine terms, each represented by an amplitude and a phase. The transformation is the casting of data for a periodic phenomenon into such an alternative form, or the reconstruction of the phenomenon from its mathematical expression, e.g. reconstruction of a molecular model of a compound from an analysis of the X-ray diffraction pattern of its crystal, which is a three-dimensional periodic structure. (*see also* phase problem)

Fourier transform-mass spectrometry (FT-MS) A variant of mass spectrometry in which a molecular ion is stabilized in a cyclotron orbit and detected by the frequency of its oscillation. The consequent long half-life and non-destructive detection account for the high sensitivity of the method. (*see also* mass spectrometry)

FPERT A variation on the phenol-emulsion reassociation technique protocol that includes formamide.

FR Framework region. (*see* complementarity-determining region)

fragment reaction assay A measure of the transferase phase of protein synthesis independent of mRNA and many ribosomal factors; i.e. the transfer of an amino acid from its position esterified to the 3′-end of a tRNA (or partial digest of a tRNA) to the N-terminus of the growing peptide chain.

• Moore, P.B. (1992) Nature (London) **357**, 439

frame-shift A mutation that throws out of register the normal reading of triplet codons during translation; usually caused by an insertion (or deletion) of one or two nucleotides into (or from) the gene.

frame-shift suppressor A mutation that circumvents the effects of a frame-shift, e.g. a mutation in the anticodon region of a tRNA that allows it to recognize a tetranucleotide codon.

framework model A proposed sequence of protein folding events in which the random coil, or denatured form, passes through three stages: the formation of *fluctuating embryos*, in which disconnected regions of secondary structure are formed; collapse of these regions into an *intermediate compact structure* or *molten globule* state that resembles the native conformation in overall outline but lacks the amino acid residue side chain interactions; and the tightening of the structure as the native conformation is assumed. This scenario is contrasted with the *subdomain model*, in which highly organized subregions of a protein are formed before their final coming together in a process similar to a conformational change of the native protein, and the now abandoned *jigsaw-puzzle model*, in which partially folded intermediates are randomly formed and then progressively assume a greater degree of organization. In any of the folding models, the native protein conformation occupies a local energy minimum which may not be the lowest possible energy; in other words, the native structure is the result of its kinetic folding pathway. The *thermodynamic model of protein structure*, on the other hand, holds that the native conformation is the most stable and is, therefore, independent of the folding pathway. (*see also* chaperone machine; hierarchical condensation; paradox of Levinthal; two-state rule)

• Baldwin, R.L. (1990) Nature (London) **346**, 409–410; Ptitsyn, O.B. (1991) FEBS Lett. **285**, 176–181

framework region (*see* complementarity-determining region)

FRAP Fluorescence recovery after photobleaching. (= fluorescence photobleaching recovery)

free base The purine or pyrimidine moiety of a nucleoside, nucleotide or nucleic acid that is not attached to the pentose or pentose phosphate moiety. (*see also* base)

free energy The form of energy that is capable of doing work at constant pressure and temperature; a concept that bridges the requirement that energy (enthalpy plus potential energy) be conserved and that the randomness (entropy) of a system spontaneously increases; expressed as J/mol (cal/mol). (*see also* second law of thermodynamics)

free radical A molecular species containing an unpaired electron, e.g. the hydroxyl ($OH^{\bullet}$), superoxide ($O_2^{\bullet-}$) and nitric oxide ($NO^{\bullet}$) radicals. (*see also* antioxidant; lipid peroxidation; oxidative stress)

freeze-etching (*see* freeze-fracture)

freeze-fracture A technique to prepare cells for scanning electron microscopy; cleavage of a solidly frozen cell, followed by *freeze-etching*, i.e. sublimation of some of the frozen water of the cell and coating the exposed subcellular structures with a film of platinum or carbon.

freezing at reactive centres of enzymes (*see* entropy effect)

French paradox The hypothesis that the lower incidence of heart disease in Mediterranean countries is due to ingestion of constituents of red wine, possibly flavinoids, that act to inhibit oxidation of low-density lipoproteins.

• Leake, D. (1995) The Biochemist **17 (1)**, 12–15

Fresco–Alberts–Doty model (= hairpin loop)

FRET Fluorescence resonance energy transfer. (= resonance energy transfer)

frictional coefficient A measure of the size and asymmetry of a molecule in solution, derived from hydrodynamic measurements, e.g. diffusion, ultracentrifugation, mobility in electrophoresis; equal to RT/ND, qE/v, and $M(1-\upsilon\rho)/NS$, where R is the gas constant, T is the absolute temperature, N is Avogadro's number, D is the diffusion coefficient, q is the macromolecule's charge, E is the electric field strength, v is the rate of electrophoretic movement, M is the molecular mass, υ is the partial specific volume, ρ is the solution density and S is the sedimentation coefficient.

FT-MS (= Fourier transform-mass spectrometry)

Fuelgen reaction A qualitative colorimetric method for identification of DNA, especially in cytochemistry; treatment with fuchsin sulphurous acid to produce a red colour.

functional cloning (*see* positional candidate approach)

functional divergence ratio (FDR) A measure of the evolutionary selective pressure on a protein. In a family of homologous proteins, or the genes encoding them, the amino acid residues are categorized as functional (e.g. those at the binding site and catalytic site) or non-functional. The FDR is the ratio of mutations found among functional residues to those found among non-functional residues.

furanose The form of a sugar when it is condensed into a five-membered ring. It consists of four carbon atoms and the oxygen atom that is the link to the anomeric carbon atom. (*see also* pyranose)

futile cycle (substrate cycle) The coupling through common substrate/product pairs of an energy-requiring enzymic reaction with another energy-producing reaction that regenerates one of the products of the first reaction. The net result is the expenditure of energy, e.g. the phosphofructokinase and fructose phosphate phosphatase reactions that together generate and hydrolyse fructose bisphosphate with the net consumption of ATP. Such a cycle is potentially useful in the rapid response of metabolism to regulatory factors.

G

GABA shunt An alternative route from α-oxoglutarate to succinyl-CoA as part of the tricarboxylic acid cycle. Rather than going through the α-oxoglutarate dehydrogenase reaction, α-oxoglutarate condenses with NH_3 to form α-iminoglutarate, then is reduced to glutamic acid, decarboxylated to γ-aminobutyrate (GABA), transaminated to succinic semialdehyde and oxidized to succinyl-CoA.

gamete A spermatozoon or ovum; a haploid germ cell.

gamete imprinting (*see* imprinting)

gamma (γ)-globulin (*see* globulin)

gap junction (cell-to-cell channel) A connection between adjacent cells that allows small molecules and ions, e.g. amino acids, sugars and nucleotides, to pass from the cytoplasm of one cell to the cytoplasm of the other.

gas-phase analogy (*see* entropy effect)

gas-phase sequencing (*see* solid-phase sequencing)

gastrulation During early embryogenesis, the invagination and reshaping of the cells of the blastoderm that results in differentiation into ectoderm, endoderm and mesoderm.

gate A regulated channel through which ions pass across the plasma membrane; *gating* is the opening and closing of the channel.

gatekeeper protein A protein that monitors transfer of a protein from the endoplasmic reticulum to the Golgi apparatus and prevents transfer of newly synthesized proteins with inappropriate conformations or with unpaired thiol groups.

gating (*see* gate)

GATT (= gene amplification with transcription/translation)

gauche conformation An arrangement in space of the carbon–carbon bonds of the alkyl chains of a fatty acid residue of a membrane lipid. The backbone of the chain is rotated 120° from the *trans* conformation, to force the backbone into making a 60° turn through the bond.

GAWTS (= genomic amplification with transcript sequencing)

G band (*see* AT queue)

GC clamp (*see* denaturing gradient gel electrophoresis)

gd-PCR (= gene dosage PCR)

gel filtration chromatography A technique for separation of soluble compounds according to size, based on their relative ability to penetrate the carefully sized pores of the stationary gel phase.

gel retardation assay (*see* gel shift assay)

gel shift assay (electrophoretic mobility shift assay; EMSA) A method to identify by function a band on a polyacrylamide gel. Electrophoresis is performed on a mixture of proteins before and after treatment with a strong ligand that is specific to one component, and note is made of the absence or shift in position of a protein due to prior binding to the ligand. A variant is the *gel retardation assay*, an approach to detecting DNA-binding proteins by their ability to bind to a short radiolabelled DNA fragment and, as a result, decrease its electrophoretic mobility in a gel.

gene In genetics, a unit inferred from the pattern of inheritance; in molecular biology, defined narrowly as a section of DNA that is expressed as RNA or, more widely, as a coding sequence of DNA and associated regulatory sequences.

gene amplification The synthesis of multiple copies of a specific gene, which remain as tandem repeats within the chromosome or are segregated as satellite DNA.

gene amplification with transcription/ translation (GATT) A method for the production of up to 10^{10} protein molecules per DNA by a continuous process in a homogeneous solution.

• Resto, E., Iida, A., Van Cleve, M.D. and Hecht, S.M. (1992) Nucleic Acids Res. **20**, 5979–5983

gene dosage PCR (gd-PCR) A method for quantification of the relative amount of a gene in a biological sample. One primer is constructed to hybridize with a gene known to be in the chromosome in question (e.g. chromosome 21 in the case of Down's syndrome) and another to hybridize to an unrelated gene in a different chromosome. Quantification of the amplified PCR products is sufficiently precise to distinguish a 50% increase in the chromosome 21 gene.

• Taylor, G.R. and Logan, W.P. (1995) Curr. Opin. Biotechnol. **6**, 24–29

gene probe (= hybridization probe)

gene product The result of gene expression, i.e. an RNA or protein molecule.

general acid In chemistry, a proton donor that can participate in catalysis. (*see also* general base; specific acid)

general base In chemistry, a proton acceptor that can participate in catalysis. (*see also* general acid; specific base)

general genetic recombination Recombination between homologous chromosomes that may occur during meiosis anywhere along their lengths. (*see also* Holliday model; Meselson–Radding model)

gene targeting The introduction of a homologous DNA sequence into a specific site in the genome of a cell, either by replacement of the former sequence, i.e. *sequence replacement*, or by insertion into the former sequence, i.e. *sequence insertion*. A vector that is homologous and co-linear with a partial sequence of the targeted gene is introduced into an appropriate cell, e.g. a stem cell of some sort; it is incorporated into the genome of some of the cell's progeny and replaces the former sequence. If, however, homologous regions of the vector hybridize to the gene, and it is then incorporated into the gene by homologous recombination, the result is a disruption of the gene by insertion of the vector sequence.

gene therapy The *in vivo* introduction of a functional gene, and its expression, in the germ line cells (*germ line therapy*) or somatic cells (*somatic gene therapy*) of an individual who does not possess the normal gene.

• Mulligan, R.C. (1993) Science **260**, 926–932

genetic code The series of codons of mRNA, each of which specifies a single specific amino acid.

genetic code blocker (= code blocker)

genetic fingerprinting A technique for establishing genetic relationships, especially in forensic medicine, by comparison of the occurrence of uncommon genetic markers.

genetic footprinting An approach to the identification of the function of a large number of putative genes of a micro-organism, such as may have been uncovered by genome sequencing. A transposable element is inserted into a large number of sites in the genome of the micro-organism, and the mutagenized population is then grown under a wide variety of conditions that may suggest a gene's function. After many population doublings under each set of conditions, the micro-organisms' DNA is extracted and used as a resource for analysis of as many of the putative genes as is desired. PCR primers are constructed to hybridize with the transposable element and with a putative gene. PCR amplification using the primer pair will produce a spectrum of bands on separation by polyacrylamide-gel electrophoresis, one from each mutation that can grow under the set of conditions, i.e. a genetic footprint of the gene. A comparison of the gene's footprints under permissive and restrictive conditions will indicate, by the absence of bands in the latter footprint, those growth conditions that require the function of the gene.

• Smith, V., Botstein, D. and Brown, P.O. (1995) Proc. Natl. Acad. Sci. U.S.A. **92**, 6479–6483

genetic map (*see* mapping)

genetic marker A DNA sequence that can be recognized and thus used to characterize the larger DNA sequence and the chromosome in which it occurs.

genome The genetic complement of an organism represented by its DNA or, in some viruses, its RNA.

genomic amplification with transcript sequencing (GAWTS) A DNA sequencing method that involves PCR amplification of the target and uses a primer with an attached phage promoter sequence, transcription to produce a single-stranded RNA and reverse transcription with dideoxy terminator nucleotides that upon electrophoresis will generate a sequencing ladder.

• Buerstedde, J.-M. and Sommer, S.S. (1991) in PCR Topics: Usage of Polymerase Chain Reaction in Genetic and Infectious Diseases (Rolfs, A., Schumacher, H.C. and Marx, P., eds.), pp. 9–14, Springer-Verlag, Berlin

genomic DNA DNA that has been isolated from a cell and therefore contains introns, as opposed to cDNA.

genomic library A collection of transformed cells, each of which contains DNA fragments; the entire population represents the total genome of an organism, e.g. a rat library containing DNA fragments which together comprise the entire rat genome. Appropriate screening methods can select a single transformed cell that contains a specific gene. (*see also* cDNA library)

genomic mismatch scanning (GMS) A method to detect and isolate DNA sequences that are candidate genes for inherited disorders for which the gene product is unknown, based on the absence of mismatches in DNA sequences between an affected individual and a heterozygous or carrier progenitor. Large DNA segments are prepared from the genomic DNA of the two related individuals in such a way (e.g. by leaving 3'-overhangs) that they will not be degraded by subsequent exonuclease digestion, e.g. by *Exo*III. The DNA from one individual is enzymically methylated and annealed with the DNA of the second, heterohybrids (two methylated or two unmethylated strands) are cleaved by appropriate endonucleases, e.g. *Dpn*I and *Mbo*I, and the uncleaved duplexes are scanned for single-base mismatches by methyl-directed mismatch repair enzymes that leave single-strand nicks that are attacked by *Exo*III. The duplexes that survive all these tests are those that are shared by the two related individuals and are therefore candidates for the affected gene.

• Nelson, S.F., McCusker, J.H., Sander, M.A., et al. (1993) Nature Genet. **4**, 11–17; Kolodner, R.D. (1995) Trends Biochem. Sci. **20**, 387–401

genomic subtraction (= subtractive DNA cloning)

genomic tag model The hypothesis that a tRNA-like loop (the tag) of the genomic RNA of some viruses that functions in copying a strand of RNA into a complementary RNA or DNA strand is a relic of the part of the RNA that functioned in initiating its own replication in the primordial RNA world.

• Gibbons, A. (1992) Science **257**, 30–31

genotype The genetic composition of a cell or organism, usually with reference to a particular set of alleles that may be homozygous or heterozygous for a trait; the *phenotype* is the appearance of the cell or individual due to actual expression of the alleles that are present.

germ cell (= gamete)

germ line therapy (*see* gene therapy)

ghost An erythrocyte plasma membrane; the product of red cell lysis.

gibberellin A type of plant hormone; a diterpene that is responsible for cell differentiation, e.g. gibberellic acid.

Gibbs–Donnan effect (= Donnan effect)

global regulator A metabolite that regulates many operons, e.g. glucose, as it regulates several anabolic reactions in a bacterium.

global regulatory circuit A series of operons that are induced together, e.g. the genes for the heat-shock proteins or those for the SOS repair proteins.

globular Descriptive of the folding of a protein upon itself in several convolutions to create interactions of the side chains in many weak salt bridges, hydrogen bonds and hydrophobic bonds that result in a roughly spherical (globular) molecular shape.

globulin Archaic nomenclature for a protein that is sparingly soluble in water but is soluble in dilute salt solutions. *Euglobulins* do not dissolve in salt-free water, whereas *pseudoglobulins* are soluble in salt-free water. *Gamma* (γ)-*globulins* are a population of globulins defined further by their electrophoretic behaviour, and include the immunoglobulins. (*see also* albumin)

G loop (*see* AT queue)

glucocorticoid An adrenal steroid that increases blood glucose concentration, e.g. cortisol.

glucogenic amino acid An amino acid whose carbon skeleton can be metabolically converted, at least in part, into glucose. (*see also* ketogenic amino acid)

gluconeogenesis The biosynthesis of glucose from smaller, non-carbohydrate, metabolites, i.e. amino acids, tricarboxylic acid cycle intermediates, lactate or glycerol.

glucose effect The repression of some bacterial enzymes by the presence of glucose in the growth medium. Depression of cyclic AMP levels due to the availability of glucose is mediated by inhibition of adenylate cyclase by a glucose catabolite. In the absence of glucose, a complex between cyclic AMP and its binding protein attaches to specific regions of DNA and activates transcription. Cyclic AMP acts as a starvation signal, as it does in animal cells, but in bacteria it acts at the transcriptional level. (*see also* diauxie; positive control)

glucose repression (*see* glucose effect)

glycation The reversible condensation of the carbonyl group of a sugar with an amino group of a protein to form a Schiff base that often triggers subsequent irreversible rearrangements, e.g. the condensation of glucose with haemoglobin A to form haemoglobin A_{1c}.

glycobiology The study, at the molecular level, of the structural and functional roles of carbohydrate-containing structures and of the interactions of carbohydrates with other structures, especially proteins. (*see also* glycoconjugate)

- Opdenakker, G., Rudd, P.M., Ponting, C.P. and Dwek, R.A. (1993) FASEB J. **7**, 1330–1337; Lee, Y.C. and Lee, R.T. (1995) Acc. Chem. Res. **28**, 321–327

glycocalyx The carbohydrate coating on the external surface of a cell membrane.

glycoconjugate A compound composed of an oligosaccharide linked to a protein or lipid, e.g. mucins (which are glycoproteins), globosides (which are *N*-acetyl-sphingosine conjugates of oligosaccharides), and the endotoxin of Gram-negative bacteria. (*see also* lipopolysaccharide)

glycoform One of several variants of a glycoprotein that have identical polypeptide moieties but differ in the site of glycosylation (site heterogeneity) and/or the nature of the oligosaccharide moiety. Glycoforms are characteristic of the tissue of origin, or of the disease or mutagenic state.

glycogen storage disease A defect in the turnover of glycogen that leads to increased amounts of the normal or abnormal polysaccharide in affected cells.

glycolipid A compound with both lipid and sugar moieties, e.g. lipopolysaccharide of Gram-negative bacteria cell walls, blood group antigens, gangliosides (carbohydrate moieties attached to N-fatty acylsphingosine), cerebrosides and ceramides.

glycolysis One of the central pathways of metabolism in most eukaryotes and many other cells; the sequence of enzymic reactions that converts glucose into lactic acid (*anaerobic glycolysis*) or into pyruvate (*aerobic glycolysis*). (*see also* Embden–Myerhof pathway)

glycoprotein A protein with one or more carbohydrate moieties attached to it.

glycoprotein hormone A representative of a class of hormones that share nearly identical A-chains and unique B-chains, each with about 125 amino acid residues and N-linked or O-linked carbohydrate moieties, e.g. thyrotropin, follitropin.

glycosaminoglycan A subunit of a proteoglycan; an oligosaccharide that contains repeating disaccharide residues. One half of each disaccharide is an aminosugar residue. Glycosaminoglycans include the sulphated chondroitins, dermatans and keratans, and the unsulphated hyaluronans. (*see also* ambidexteran; core protein; link protein; proteodermochondran sulphate)

glycoside A metabolite formed by conjugation with a sugar through its anomeric carbon atom, e.g. β-methyl glucoside. (*see also* aglycone)

glycosidic bond The linkage of a sugar hemiacetal or hemiketal through its anomeric carbon with another moiety. (*see also* glycoside)

glycosylation Post-transcriptional modification of a protein by the addition of a carbohydrate moiety. (*see also* N-linked carbohydrate; O-linked carbohydrate)

glyoxylate cycle (Krebs–Kornberg cycle) A series of metabolic reactions in plants and bacteria, the net result of which is production of a 4-carbon compound from two 2-carbon compounds. The unique reactions are catalysed by malate synthase and isocitrate lyase.

glyoxysome An organelle in plant cells in which the glyoxylate cycle operates.

glypiated Descriptive of a glycosylphosphatidylinositol (GPI)-anchored protein, i.e. one that is associated with a cell membrane through an attached GPI group. An alternative term is *piglylated*, from phosphatidylinositol glycan (PIG).
- Dotti, C.G., Parton, R.G. and Simons, K. (1991) Nature (London) **349**, 155–161

glypican (*see* proteoglycan)

GMS (= genomic mismatch scanning)

Golgi apparatus A stack of flattened vesicles that functions in the post-translational processing and sorting of proteins. The Golgi receives proteins from the rough endoplasmic reticulum (RER) and directs them to secretory vesicles, lysosomes or the cell's plasma membrane. The movement of the proteins takes place by transfer vesicles that bud off from the RER or Golgi and fuse with the Golgi, lysosomes or plasma membrane. The surface of the Golgi that faces the RER is termed the *cis*-Golgi, the surface towards the apical membrane is the *trans*-Golgi, and the intermediary part is the medial Golgi.

Gortner and Grendel model A model in which erythrocyte membranes comprise only phosphoacylglycerols, i.e. no protein or sterol component is envisaged; the model presents a bilayer in which each leaflet is composed of lipids stacked perpendicular to the plane of the membrane, with their alkyl chains oriented inwards (towards the opposing leaflet) and with the polar groups oriented towards the exposed surfaces. (*see also* Danielli–Davson model; fluid mosaic model; unit membrane Danielli–Davson model)

G₁ period (*see* cell cycle)

G₂ period (*see* cell cycle)

GPI Glycosylphosphatidylinositol. (*see* glypiated)

G-protein GTP-binding protein; one of a superfamily of proteins that function in, for example, signal transduction (e.g. the G-protein associated with the β-adrenergic receptor), polymerization (e.g. tubulin), ribosomal protein synthesis (e.g. the translocase), cell differentiation (ras proteins) and intracellular transport of proteins, vesicles or cytoskeletal elements (e.g. dynamin). The common functional feature of the family is an affinity for a target protein in the presence of GTP which is lost upon hydrolysis to GDP.

• Bourne, H.R., Sanders, D.A. and McCormick, F. (1990) Nature (London) **348**, 125–132; Hilgenfeld, R. (1995) Curr. Opin. Struct. Biol. **5**, 810–817

G-quartet (= G-tetrad)

Gram-negative Descriptive of bacteria that do not stain by the Gram method, i.e. bacteria with two membranes. (*see also* Gram-positive; lipopolysaccharide)

Gram-positive Descriptive of bacteria that stain by the Gram method, i.e. bacteria with single membranes. (*see also* Gram-negative)

granulocyte A granular leucocyte; a basophil, eosinophil or polymorphonuclear leucocyte (neutrophil).

granum A closely packed stack of thylakoid membranes in a chloroplast.

granzyme A T-cell (granulocyte) proteinase.

Greek key A topology found in some protein structures; antiparallel β-strands are organized into a barrel in which strands that are adjacent in the primary structure are not always adjacent in super-secondary structure but, when represented in two dimensions, form a Greek key pattern.

gRNA (= guide RNA)

ground state The unexcited electronic state. (*see also* singlet state; triplet state)

ground substance The histologically featureless extracellular matrix of connective tissue, largely composed of proteoglycans.

group I self-splicing (*see* self-splicing)

group II self-splicing (*see* self-splicing)

group transfer hydration (*see* entropy effect)

group transfer potential A quantitative measure of the strength of attachment of a moiety (the group) to the rest of the molecule; the difference in standard free energies between the group attached to the molecule and the group attached, usually, to water, e.g. the standard free energy $(\Delta G^{\circ\prime})$ of the hydrolysis of ATP to ADP and phosphate, -30.5 kJ/mol $(-7.3$ kcal/mol).

growth factor A protein that binds to receptors on specific cells and promotes their growth, e.g. nerve growth factor, epidermal growth factor, platelet-derived growth factor.

G-tetrad (G-quartet) A structure that is proposed to stabilize via Hoogstein base pairings the interaction of ends of two chromosomes. From each double-stranded (ds) DNA extends a 3′-end beyond the complementary 5'-end, which creates an overhang of a TTGGGGTTGGGG sequence. The overhang of each dsDNA is bent into a hairpin loop in which the guanines form the Hoogstein base pairs with themselves. They are also hydrogen bonded to identical hairpin loops of the other dsDNA, thus forming a stack of four planes in each of which four guanines are associated in a hydrogen-bonded network.

• Sundquist, D.I. and Klug, A. (1989) Nature (London) **342**, 825–829

GTPase protein (= G-protein)

GTP-binding protein (= G-protein)

guide RNA (gRNA) A transcript of maxicircle or minicircle DNA that in lower eukaryotes is thought to function in *RNA editing*. Thus by the addition or deletion of U residues at specific sites in the coding regions of mRNA, itself derived from maxicircle DNA as an untranslatable *cryptogene*, a translatable open reading frame is created. The gRNA attaches to an unedited region (*pre-edited region;*

PER) which at a GA-rich sequence anchors the oligo(U) 3'-end of the gRNA and at a downstream sequence anchors the complementary 5'-end of the gRNA. Between the two anchored regions of the gRNA, a non-complementary oligonucleotide sequence loops out. An endonuclease cleaves the PER between the two anchors and uses a G residue of the gRNA loop as a template to patch in a U residue. The gRNA is then free to migrate to another site for editing. In some cases this mechanism inserts nearly half the U residues necessary for translation of an mRNA. This extensive editing of a single mRNA is termed *pan-editing*.

• Simpson, L. (1990) Science **250**, 512–513; Feagin, J.E. (1990) J. Biol. Chem. **265**, 19373–19376

guide sequence In the self-splicing of RNA, the purine-rich region of the intron that binds a pyrimidine-rich region of the upstream exon to the site where transesterification takes place.

gynogenic (*see* anthrogenic)

H

haemoglobinopathy (hemoglobinopathy)
An abnormality that may produce a disease the basis of which is an alteration in haemoglobin structure or synthesis.

haemolysis (hemolysis) The bursting of erythrocytes due, for example, to a hypotonic environment.

haemostasis (hemostasis) The stopping of blood loss from the vascular system by vasoconstriction, formation of a platelet plug and blood clotting.

hairpin bend (= reverse turn)

hairpin loop (Fresco–Alberts–Doty model)
A region in a single polynucleotide strand containing adjacent complementary sequences that allows the polynucleotide to fold back upon itself and form a double-stranded section, as in the cloverleaf structure of a tRNA molecule.

hairpin RNA (*see* ribozyme)

half-chair conformation The conformation of a pyranose ring that has been strained to place four adjacent atoms in one plane, e.g. the postulated transition state of the *N*-acetylglucosamine residue during lysozyme catalysis, in which C-1 becomes a planar carbocation.

half-life When a property (e.g. radioactivity, enzyme activity, etc.) decreases at a rate proportional to its concentration, the time, $t_{1/2}$, taken for the property to decrease to one-half its initial value; related to the first-order rate constant, k, by $t_{1/2} = \ln 2 / k$.

half-of-the-sites activity An extreme form of negative co-operativity shown by a multisubunit enzyme in which, as a substrate binds to the active site of one subunit, the affinity for the substrate of an adjacent subunit decreases.

half site One of the two symmetrical duplex moieties of a palindromic site in DNA.

hammerhead RNA (*see* ribozyme)

handshake model (*see* induced-fit theory)

hanging-drop technique A method for slowly changing concentrations in a protein solution in an attempt to prepare crystals. A small volume of protein solution (of the order of 5 μl) on a microscope coverslip is sealed drop-side down over a small well containing a larger volume (of the order of 1 ml) so that, over time, volatile solvents will exchange between the hanging drop and the solution in the well via the vapour phase and concentrate the protein solution or allow alcohol to pass over to it. The progress of crystallization is observable in the droplet through the coverslip.

haploid Descriptive of the number of chromosomes of a gamete; one of each pair of chromatids. (*see also* diploid)

hapten A small molecule which by itself is not an antigen, but which as a moiety of a larger structure (a *haptenic determinant*) can serve as an antigenic determinant.

haptenic determinant (*see* hapten)

Harden–Young ester Obsolete name for fructose 1,6-bisphosphate

Hatch–Slack pathway (= C_4 photosynthesis)

HAT medium (= hypoxanthine/aminopterin/thymidine medium)

H⁺:ATP ratio (*see* P:O ratio)

Haworth projection A two-dimensional representation of pyranose and furanose structures in which they are shown as hexagons and pentagons respectively. The lower ring atoms are understood as being towards and the upper ones away from the observer; the substituents of ring carbon atoms are shown directly above and below the apices of the polygon. (*see also* Fischer convention)

HCA (= hydrophobic cluster analysis)

headward Descriptive of a mechanism of polymerization. The head of a biosynthetic monomer is defined as the chemically activated site (i.e. the carbonyl groups of amino acyl-tRNA or fatty acyl-CoA esters, the hemiacetal oxygen of a UDP-monosaccharide, the α-phosphate of a nucleoside triphosphate, the alkyl oxygen of isopentenyl pyrophosphate), and the tail as the other site on a monomer that will be joined to an adjacent moiety. Headward elongation occurs when a monomer condenses by displacement of the activating group (tRNA, CoA, pyrophosphate) of the growing polymer (protein, fatty acid or terpene biosynthesis). Conversely, *tailward* elongation occurs when the condensation removes the activating group of the newly added monomer (nucleic acid or polysaccharide biosynthesis).

heat-shock consensus element (HSE) A 14 base pair polynucleotide sequence upstream from structural genes for heat-shock proteins, required for optimal expression. A somewhat longer polynucleotide sequence that includes flanking regions of the HSE is the *heat-shock regulatory element.*

heat-shock protein One of the proteins produced by some cells when they are stressed, e.g. by an abrupt increase in temperature.

heat-shock regulatory element (*see* heat-shock consensus element)

heavy chain One of two or more kinds of polypeptide chain of a heteromultimeric protein, distinguished by molecular mass, e.g. one of the larger polypeptide chains that, when combined with another heavy chain and two light chains, make up an immunoglobulin molecule; also the largest of the polypeptide chains of any multimeric protein with non-identical subunits.

heavy-chain library (*see* repertoire cloning)

heavy-ion-induced desorption (*see* mass spectrometry)

heavy subunit The larger of the two ribonucleoprotein complexes that make up a ribosome; more generally, the largest of the subunits of any complex. (*see also* light subunit)

Heinz body A precipitate in the cytoplasm of an erythrocyte of a spontaneously oxidized, unstable, haemoglobin. (*see also* inclusion body)

helical wheel A representation of the array of amino acid residue side chains around an α-helix viewed end-on, down the axis of the helix. (*see also* Eisenberg plot; hydrophobic moment)

helix–coil transition In nucleic acid and protein chemistry, the melting, or co-operative thermal breakdown of the hydrogen bonds that stabilize the secondary structure, of the macromolecule.

helix–loop–helix (HLH) A motif found in some DNA-binding proteins.

helix–turn–helix (HTH) motif A motif found in many sequence-specific DNA-binding proteins, e.g. in the lambda repressor and related proteins. (*see also* HLH protein)

helper T-cell (*see* T-cell)

hemiacetal The product of reversible condensation of an aldehyde and an alcohol in which the alcoholic hydroxy group adds across the carbonyl group of the aldehyde, e.g. glucose acting as both an aldehyde and an alcohol to give the internally condensed glucopyranose; also the bond formed by this condensation. An *acetal* is the product formed by abstraction of the hydroxy group of a hemiacetal and the hydrogen of a second alcohol, e.g. a glucopyranoside or a polysaccharide. (*see also* hemiketal)

hemicatenane (*see* plectonemic)

hemi-channel (connexon) Half of a gap junction. Being embedded in the membrane of each of the cells and composed of six molecules of the integral protein connexin, the hemi-channel allows communication between the cytoplasm of adjacent cells.

hemidesmosome A structure of the plasma membrane of an epithelial cell that extends into both the cytoplasm and the extracellular space, and anchors the cell to the extracellular matrix of the basal lamina.

hemiketal The product of reversible condensation of a ketone and an alcohol in which the alcoholic hydroxy group adds across the carbonyl group of the ketone, e.g. fructose acting as both a ketone and an alcohol to give the internally condensed fructofuranose; also the bond

formed by this condensation. A *ketal* is the product formed by abstraction of the hydroxy group of a hemiketal and the hydrogen of a second alcohol, e.g. a fructofuranoside. (*see also* hemiacetal)

hemisubstrate An enzyme inhibitor that acts in the initial phases of enzyme action as a substrate, but does not complete the catalytic cycle to regenerate an active enzyme; e.g. organophosphate ester anti-cholinesterases, which mimic acetylcholine in that the phosphoric acid moiety of the hemisubstrate acts like the acetly group of acetylcholine to form an ester with the enzyme's active-site serine residue but, unlike the acetyl-enzyme intermediate of the normal enzymic action, the phosphoryl-enzyme is only very slowly hydrolysed to regenerate an active enzyme.

hemizygous Descriptive of a trait that is carried on the X chromosome and is therefore present in the male population at half the frequency that occurs in females.

hemolysis (= haemolysis)

hemoglobinopathy (= haemoglobinopathy)

hemostasis (= haemostasis)

Henderson–Hasselbalch equation A logarithmic form of the equation for the dissociation of a weak acid: $pH = pK_a + \log[\text{salt form}]/[\text{acid form}]$.

hepato- A prefix that indicates the liver, e.g. hepatocyte.

hetero- (*see* homo-)

heterochromatin In cytology, the heavily staining regions of chromosomes that contain condensed chromatin. (*see also* euchromatin)

heteroduplex A partially double-stranded polynucleotide in which one strand contains sequences not fully complementary to the opposite strand, e.g. mRNA, which contains no introns, annealed to the coding strand of its corresponding gene, which contains introns. Heteroduplexes are products of recombination between DNA duplexes at a region of heterology. (*see also* D-loop)

heteroduplex analysis (mismatch analysis) A method for detection of single base substitutions in DNA fragments. The PCR-amplified putatively mutated DNA fragment is mixed with homologous normal DNA fragments; any mutated sequence is much more dilute than the normal sequence and will therefore be less likely to re-anneal. They will appear on polyacrylamide-gel electrophoresis as band mobility shifts, especially under mildly denaturing conditions, and will indicate the presence of heteroduplexes.
- White, M.B., Carvalho, M., Derse, D., et al. (1992) Genomics **12**, 301–306

heteroduplex mapping An electron microscopic technique that compares sequence relationships of two polynucleotides, e.g. a spliced eukaryotic mRNA with a coding strand of DNA. Introns of the DNA strand appear as D-loops that deviate from the base-paired sections where the two strands are complementary.

heterogeneous nuclear RNA (= pre-mRNA)

heterokaryon A hybrid of cells from different species, e.g. a fusion of a human and a mouse cell.

heterolactate fermentation (*see* homolactate fermentation)

heterolytic cleavage The splitting of a covalent bond that leaves both bonding electrons with one of the atoms. (*see also* homolytic cleavage)

heteromeric (*see* homomeric)

heterophagy The action of a lysosome of a phagocytic cell to digest extracellular materials. (*see also* autophagy)

heteropolysaccharide A carbohydrate composed of more than one kind of sugar monomer.

heterotrophic In microbiology, descriptive of organisms that require organic material for growth. (*see also* autotrophic; mixotrophic)

heterotropic enzyme An enzyme that is controlled by an allosteric activator that is not also a substrate. (*see also* homotropic enzyme)

heterozygote (*see* allele)

hexose monophosphate shunt (= pentose phosphate pathway)

hierarchical condensation The hypothesis that, during the folding of a protein in solution, some regions are much more likely to adopt a secondary structure and that, once formed, these interact to direct the subsequent course of folding. This hypothesis is a component of the framework model.

high-energy bond A chemical bond whose hydrolysis results in the generation of ≥ 30 kJ (7 kcal) of energy or, if coupled to an energetically unfavourable reaction, can drive that reaction forward.

high-energy phosphate A phosphate anhydride, enol phosphate or similar compound whose hydrolysis is associated with a large decrease in free energy. (*see also* high-energy bond; phosphagen)

high-mannose-type carbohydrate One type of glycoprotein moiety that is attached to the β-amide nitrogen of an asparagine residue (N-linked), and contains exclusively mannose residues at the non-reducing ends. (*see also* complex-type carbohydrate; hybrid-type carbohydrate)

high-mobility-group (HMG) protein One of a class of non-histone nuclear nucleoproteins, i.e. acidic proteins of chromatin that are characterized by their elution properties from an ion-exchange column.

high-performance liquid chromatography (= high-pressure liquid chromatography)

high-pressure (high-performance) liquid chromatography (HPLC) A technique for rapid separation of solutes on a solid support, based on ion-exchange, gel permeation or partition principles; *reverse-phase HPLC* is when the stationary phase has a non-polar coat; *microbore* or *narrow-bore HPLC* is adapted to small quantities and high sensitivity by scaling down.

high-voltage electrical discharge (= electropermeabilization)

HIID Heavy-ion-induced desorption. (*see* mass spectrometry)

Hill coefficient A measure of co-operativity, i.e. the number h in the equation $[HbO_2]/[Hb] = k(pO_2)^h$ that empirically describes the dependence of haemoglobin (Hb) oxygenation on the partial pressure of oxygen (pO_2).

Hill plot A graphical representation of binding data, especially for oxygen binding to haemoglobin; a plot of $\log [Y/(1-Y)]$ against $\log pO_2$, where Y is the fraction of binding sites occupied and pO_2 is the partial pressure of oxygen; the slope of the plot is the Hill coefficient (h).

Hill reaction The concomitant reduction of an electron carrier and oxidation of water to molecular oxygen that is performed by illuminated green plants; the *light reaction* of photosynthesis. (*see also* Calvin cycle; Z-scheme)

hinge A flexible polypeptide sequence connecting two domains of a protein that can move with respect to each other.
  • Dobson, C.M. (1990) Nature (London) **348**, 198–199

Hirt lysate A preparation from which low-molecular-mass extrachromosomal DNA may be isolated; made by rendering insoluble the higher-molecular-mass DNA and other cellular structures by treatment of cells with a detergent and a concentrated salt solution.

histone One of five small, basic, proteins that are incorporated into chromatin. (*see also* nucleosome)

HLA (= human leucocyte antigen)

HLH (= helix–loop–helix)

HLH protein One of a group of proteins defined by their characteristic topography (the helix–loop–helix). These proteins form DNA-binding heterodimers composed of two proteins, one from a ubiquitous and one from a tissue-specific class of HLH proteins. The heterodimers are involved in regulation of gene expression. (*see also* helix–turn–helix motif)

HMG protein (= high-mobility-group protein)

hnRNA Heterogeneous nuclear RNA. (= pre-mRNA)

Hofmeister series (= lyotropic series)

Hogness box (= TATA box)

Holliday junction (*see* double-strand-break repair model; Holliday model; Meselson–Radding model)

Holliday model A conceptualization of general genetic recombination of newly replicated DNA during meiosis. An endonuclease nicks one strand of each of the two daughter duplexes at homologous sites which permits them to cross-over before being ligated. *Branch migration* of the crossing-over point, the *Holliday junction*, along the strands is also possible before they separate to form the recombinant products; *chi-forms* (χ-shaped structures) are generated as the strands separate. In replication of closed circular (cc)DNA, *figure-of-eight forms* are generated as the two daughter ccDNAs separate, except where they have crossed-over and still adhere to each other. (*see also* double-strand-break repair model; Meselson–Radding model)

• Strauss, B.S. (1992) Chemtracts Biochem. Mol. Biol. **3**, 40–44; Shinagawa, H. and Iwasaki, H. (1996) Trends Biochem. Sci. **21**, 107–111

holocrine Descriptive of a secretion mechanism in which a cell releases its product by rupture and death of the entire cell. (*see also* apocrine; exocrine)

holoprotein The complex of an apoprotein with its prosthetic group or metal ion.

homeobox A highly conserved 180-base polynucleotide sequence that controls body part-, organ- or tissue-specific gene expression, e.g. development of antennae or legs of *Drosophila*, but also in a wide variety of other eukaryotes. It codes for a helix–turn–helix DNA-binding region, a *homeodomain*, of proteins that are transcription factors.

• Edelman, G.M. and Jones, F.S. (1992) Trends Biochem. Sci. **17**, 228–232; Kornberg, T.B. (1993) J. Biol. Chem. **268**, 26813–26816; White, R. (1994) Curr. Biol. **4**, 48–50

homeodomain (*see* homeobox)

homeostasis (= homoeostasis)

homing The migration of lymphocytes to areas of inflammation. (*see also* intron homing)

homo- A prefix that signifies an entity, especially a polymer, of uniform subunits, e.g. glycogen, a homopolymer of α-D-glucose; contrasted with *hetero-*, a prefix that signifies a polymer of unlike components, e.g. hyaluronic acid. Also a prefix that signifies a product formed by the addition of a -CH_2- group, e.g. homoserine or D-homo-steroids, in which the D ring has been expanded to include six carbon atoms.

homochromatography A form of displacement chromatography in which unlabelled species, e.g. oligonucleotides, displace labelled ones to achieve improved separation. After electrophoresis in one dimension the initially separated material is transferred from a paper strip to a thin layer chromatography plate that is developed with a solution that contains the unlabelled species and displaces the labelled ones from sites on the ion-exchange stationary phase.

homoeostasis Self-regulation by an organism to maintain a relatively constant internal environment.

homogenate An unfractionated, cell-free, preparation of tissue prepared by disruption of tissue structure and breakage of cell walls, e.g. with a Potter–Elvehjem homogenizer or by use of ultrasonic vibrations.

homoiothermic Descriptive of an organism that can maintain a fixed internal temperature. (*see also* poikilothermic)

homolactate fermentation The fermentation of a hexose that produces only lactic acid, as opposed to *heterolactate fermentation*, which also produces other products.

homologous ligation A method for fusion of two oligonucleotides, independently of any restriction site they share. A region flanking the desired ligation site on one fragment is introduced into the second fragment by PCR as part of one of the primers. The first fragment and the amplified second fragment are introduced in tandem into an appropriate plasmid that can be cleaved by a restriction nuclease engineered between the two fragments. The linearized DNA is used to transform a bacterium that has a high frequency of intramolecular homologous recombinations. Subsequent homologous recombination fuses the two fragments at their homologous region.

• Kawaguchi, S.-I. and Kuramitsu, S. (1994) Trends Genet. **10**, 420

homologous recombination The non-symmetrical crossing-over of genes at homologous sequences (e.g. at *Alu* sequences of introns between different exons of a single gene) which gives rise to two mutant alleles, one a deletion and the other an insertion; recognized in haemo-globin and low-density-lipoprotein receptor mutations. (*see also* illegitimate recombination)

homologous recombination hot spot A DNA sequence that, by providing a specific protein- and RNA-binding site, shows a high frequency of homologous recombination events.

homologous replacement A method to alter a cell's genome by introduction of a vector into the nucleus of a targeted germ or somatic cell, where it can pair with the cognate sequence of the chromosomal DNA and replace the homologous region.

homologue cloning A strategy for selection of polynucleotide sequences closely homologous to one that is known by its partial hybridization with a probe designed for the latter; e.g. a labelled probe that is complementary to a polynucleotide sequence that encodes one histamine receptor may hybridize with a homologous sequence that encodes another histamine receptor under non-stringent hybridizing conditions and allow its identification and selection. (*see also* stringency)

homologue-scanning mutagenesis A strategy for identification of receptor-binding regions of a protein by substitution of analogous regions from homologous proteins in order to preserve the native three-dimensional structure of the original protein; e.g. the substitution of regions of human growth hormone with regions from pig growth hormone, human prolactin or human placental lactogen, followed by determination of binding constants for the constructs.

homology The similarity in base sequences of genes or amino acid sequences of proteins that denotes a common evolutionary origin; also the similarity of structure or function of proteins that is due to a common evolutionary origin.

homolytic cleavage The splitting of a covalent bond that leaves one of the bonding electrons with each of the atoms, thus generating free radicals. (*see also* heterolytic cleavage)

homomeric Descriptive of a complex of a single kind of subunit, e.g. bacterial glutamine synthetase; contrasted with *heteromeric*, which is descriptive of a complex of two or more kinds of subunit, e.g. haemoglobin A.

homoplasy In phylogenetic taxonomy, a non-inherited characteristic shared by a group that itself may have a common ancestor.

homopolymer tailing (= tailing)

homotropic enzyme An enzyme that is controlled by an allosteric activator that is also a substrate. (*see also* heterotropic enzyme)

homozygote (*see* allele)

Hoogstein base pairs The unique pairing that occurs when a homopyrimidine strand, which lies in the major groove of double-stranded DNA, interacts with a sequence of a Watson–Crick base-paired double helix, resulting in a triple helix. Such pairing is possible when the original double helix features a homopurine-to-homopyrimidine base-paired sequence. Allowed hydrogen-bonded base pairs are thymine to an adenine–thymine pair, or cytosine to a guanine–cytosine pair.

hook region A normally protected but reactive sequence of the α_2-macroglobulin molecule, which is a scavenger for many proteinases in the blood. A proteinase can cleave a peptide bond in the protein's bait region, but then becomes susceptible to reaction with the thioester bond formed between a γ-carboxy group and a β-thiol group. Once caught by this hook, the proteinase remains covalently bound to the protein, and is cleared with it from the circulation.

hopanoid One of a class of pentacyclic triterpenes derived from squalene that occur in the membranes of bacteria and some plants, fulfilling the functions served by sterols in the membranes of other organisms.

H⁺:O ratio (*see* P:O ratio)

hormone A substance that mediates interactions between non-contiguous cells; classically, a substance produced in minute amounts by an organ (an endocrine organ) in one part of the body and transported by the blood to a distant target organ, which it stimulates; less rigorously applied to substances that share some part of the classical description.

host–guest protocol An experiment designed to evaluate the contribution of different amino acids at a single sequence position to the stability of a polypeptide. Site-directed mutagenesis or synthetic chemical techniques are used to construct polypeptides that differ only at a single sequence position. Quantification of the stability of the polypeptide measures the contribution of each possible amino acid. (*see also* propensity)

- Swindles, M.B., MacArthur, M.W. and Thornton, J.M. (1995) Nature Struct. Biol. **2**, 596–603

hot spot A region of a polynucleotide that experiences a high frequency of mutation. (*see also* homologous recombination hot spot)

hot start PCR A variation of the basic PCR procedure that prevents copying of non-specific primer–template complexes, which may be stable at moderate temperatures, by addition of an essential reaction component only when the temperature has reached 95°C during the first cycle.
- D'Aquila, R.T., Bechtel, L.J., Videler, J.A., et al. (1991) Nucleic Acids Res. **19**, 3749

HOT technique (= hydroxylamine and osmium tetroxide technique)

housekeeping Descriptive of a gene or enzyme that serves an essential function common to all or most cells, e.g. production of energy; as distinguished from specialized functions, such as differentiation or apoptosis.

Hpall tiny fragment (= HTF island)

HPLC (= high-performance or high-pressure liquid chromatography)

HSE (= heat-shock consensus element)

HTF island A tiny fragment, 1000–2000 bp long, generated by action of the restriction nuclease *Hpa*II; usually found associated with expressed genes; characterized by the relatively rare CpG dinucleotide that occurs unmethylated.

HTH (= helix–turn–helix)

humanized antibody A chimaeric immunoglobulin of human and non-human origin. In order to minimize the possibility of the chimaera being antigenic in humans, only the hypervariable regions of a non-human monoclonal antibody, which are responsible for its specificity, replace the homologous regions of a human immunoglobulin.

human leucocyte antigen (HLA) An antigen on the surface of human cells that consists of a generally invariant β_2-microglobulin and a larger polypeptide chain that is characteristic of the individual; responsible for most cases of organ transplant rejection. (*see also* major histocompatibility complex)

Hunter and Downs plot A graphical method for determination of dissociation constants of enzyme inhibitors. Where I is total inhibitor concentration and α is the ratio of inhibited to non-inhibited rates, a plot of $I\alpha/(1-\alpha)$ on the ordinate (*y*-axis) against substrate concentration on the abscissa (*x*-axis) will give a line that intersects the ordinate at K_i; if inhibition is competitive, the slope of the plot is K_i/K_m, if non-competitive, the slope is zero. The advantage of this method lies in its not requiring that data be collected at constant substrate or inhibitor concentration. (*see also* Cornish-Bowden plot; Dixon plot; Easson and Stedman plot)

hybrid In nucleic acid chemistry, a double-stranded polynucleotide, one strand being an RNA and the other a DNA.

hybrid hybridoma A cell that produces antibodies with dual specificity because it is a fusion of two hybridomas, and therefore produces immunoglobulins characteristic of each parent cell and hybrids that display one binding site of each parental type.

hybridization (= annealing)

hybridization probe A polynucleotide, often radiolabelled, used to detect complementary sequences, e.g. an mRNA used to locate its gene by a corresponding Southern blotting method.

hybridoma A cell that is useful in production of quantities of a monoclonal antibody; a fusion of a myeloma cell and a spleen lymphocyte that, like the myeloma cell, can divide in culture and, like the spleen cell, can produce antibodies. The use of HAT (hypoxanthine/aminopterin/thymidine) medium allows selection of hybrids of myeloma cell mutants that lack hypoxanthine–guanine phosphoribosyltransferase and consequently cannot utilize hypoxanthine, with spleen cells, which do not proliferate *in vitro*. (*see also* hypoxanthine/aminopterin/thymidine medium)

hybrid-type carbohydrate One type of glycoprotein moiety that is attached to the β-amide nitrogen of an asparagine residue (N-linked) and contains both mannose and sialic acid residues at the non-reducing ends. (*see also* complex-type carbohydrate; high-mannose-type carbohydrate)

hydrazinolysis In determination of the C-terminal amino acid of a protein, cleavage of a peptide bond by insertion across it of the elements of hydrazine; in the Maxam–Gilbert method for sequencing polynucleotides, the use of hydrazine to convert pyrimidine bases into pyrazoles and remove them from the nucleic acid backbone.

hydrodynamic volume The volume of a macromolecule as deduced from its behaviour in solution. (*see also* Stokes radius)

hydrogenation The reverse of dehydrogenation.

hydrogen bond An important cohesive force of biological macromolecules, notably interstrand interactions of bases in the Watson–Crick model of DNA and of amide groups in the α-helix of proteins. The bond is the attraction of a weak acid to a weak base due to the sharing of a proton between them, e.g. -OH⋯O=, >NH⋯O=. The sharing may be considered to be an equilibrium between two tautomeric forms, e.g. -O⁻⋯HO⁺=, each form lying in an energy well with an energy barrier between them (the *double-well hydrogen bond*). If water is excluded from the area, and if the base strengths of the two atoms that share the proton are equal, the distance between the two nucleophilic atoms is shorter and the bond is much stronger [>100 kJ/mol (24 kcal/mol); the *single-well hydrogen bond*]. An intermediate case pertains [50–100 kJ/mol (12–24 kcal/mol); the *low-barrier hydrogen bond*] when the pK_a values of the nucleophiles are similar and water is largely excluded, as at the active site of an enzyme in the transition state with its substrate.

• Cleland, W.W. and Kreevoy, M.M. (1994) Science **264**, 1887–1890; Frey, P.A., Whitt, S.A. and Tobin, J.B. (1994) Science **264**, 1927–1930

hydrolase One of the class of enzymes that transfer a chemical group from donor substrates to water, e.g. peptidase, nuclease, glycosylase.

hydron A ^{1}H, ^{2}H (deuterium) or ^{3}H (tritium) atom.

hydropathy index A measure of polarity of an amino acid residue; the free energy of transfer of the residue from a medium of low dielectric constant to water. (*see also* optimal matching hydrophobicity)

hydropathy plot A graph that shows regions of hydrophobicity or polarity of a protein; a line graph that displays the hydropathy index against the position of each amino acid residue of the protein. (*see also* optimal matching hydrophobicity)

hydrophilic residue An amino acid residue that has a charged or polar side chain.

hydrophobic bond The association of non-polar side chains of a protein that is driven by minimization of the relatively unfavourable interactions of water molecules with the non-polar groups, and maximization of the favourable interaction of water molecules with themselves when they are removed from contact with the non-polar residues and are added to the bulk water phase.

• Dill, K.A. (1990) Science **250**, 297–298

hydrophobic cluster analysis (HCA) A method for comparison of amino acid sequences of proteins to detect regions of conformational similarity. Groups of neighbouring hydrophobic amino acid residues are identified in a HCA plot that is constructed by conceptually folding the entire backbone into an α-helix, rolling the helix two complete revolutions across a two-dimensional surface and noting the positions where α-carbons have contacted the surface. Each contact point is indicated according to the residues attached to it. The two-dimensional visual pattern of neighbouring hydrophobic residues, whether adjacent in the primary sequence or brought together by folding of the helix, is characteristic of the actual conformation, and serves as a basis for comparison of proteins. (*see also* structural profile)

hydrophobicity (*see* hydrophobic moment)

hydrophobicity scale A ranking of amino acid residues or side chains according to some quantitative measure of their hydrophobicity, e.g. the free energy of transfer of the side chain from dioxane or an alcohol to water.

hydrophobic moment A measure of the amphipathicity of an α-helix. Each amino acid side chain is assigned a value, positive or negative, according to its hydrophobicity. These values are used to weight vectors for each residue as they are displayed around the helix. The summation of the vectors is the hydrophobic moment, $<\mu_M>$. Mean *hydrophobicity*, $<H>$ is the sum of the hydrophobicity values divided by the number of residues. (*see also* Eisenberg plot)

hydrophobic zipper A feature of the structure of some proteins, in which neighbouring hydrophobic residues of parallel or antiparallel β-strands interdigitate and form a hydrophobic cluster.

• Gronenborn, A.M. and Clore, C.M. (1994) Science **263**, 536

hydroxylamine and osmium tetroxide (HOT) technique (chemical mismatch detection) A method for detection of single base substitutions in DNA fragments. Heteroduplexes of normal DNA and homologous mutant DNA are generated; the presence of C mismatches allows cleavage with hydroxylamine and the presence of C and T mismatches allows cleavage with osmium tetroxide. Cleavages are detected after polyacryl-amide-gel electrophoresis.

hydroxyl radical footprinting (*see* footprinting)

hyperchromic effect The increase in absorbance, frequently measured at 260 nm, when a native polynucleotide is denatured. The ratio of absorbance of the native polymer to that of the denatured or hydrolysed material is the *hyperchromicity ratio*, and ranges from 1.0 to 1.4. (*see also* hypochromicity)

hyperchromicity ratio (*see* hyperchromic effect)

hyperfine splitting The detail of a nuclear magnetic resonance spectrum, due to interactions of adjacent nuclei, that is superimposed on the Zeeman splitting.

hyperplasia An excessive growth of an organ due to an increase in the number of cells. (*see also* hypertrophy)

hypertensive Descriptive of the action of a substance in raising blood pressure. (*see also* hypotensive)

hypertonic (*see* osmolarity)

hypertrophy An excessive development of an organ, especially due to overstimulation, but without an increase in cell number. (*see also* atrophy; hyperplasia)

hypervariable region A localized sequence of a variable region of an immunoglobu-lin that shows a particularly large variation in amino acid sequence compared with all other immunoglobulins; presumably the part of the immunoglobulin that interacts directly with the antigen. (*see also* constant region; variable region)

hypochromicity The decrease in absorbance at 260 nm observed when a strand of single-stranded DNA base-pairs with its complementary strand to form double-stranded DNA. (*see also* hyperchromic effect)

hypoglycaemic (hypoglycemic) effect The action of insulin in lowering blood glucose levels by making adipocytes and skeletal muscle cells permeable to the sugar.

hypotensive Descriptive of the action of a substance in lowering blood pressure. (*see also* hypertensive)

hypotonic (*see* osmolarity)

hypoxanthine/aminopterin/thymidine (HAT) medium A growth medium that is used to isolate monoclonal-antibody-producing cells, i.e. hybrids made from fusion of myeloma cell mutants that lack hypoxanthine–guanine phosphoribosyl-transferase, and consequently cannot uti-lize hypoxanthine, with spleen cells, which do not proliferate *in vitro*. (*see also* hybridoma)

H-zone The central area of a sarcomere, identified microscopically; traversed by thick filaments but not thin filaments.

I

idiotype A classification of immunoglobulin molecules according to the antigenicity of the variable regions. Each idiotype is unique to a particular immunoglobulin raised to a particular antigen. (*see also* allotype; anti-ergotypic; isotype)

IEF (= isoelectric focusing)

illegitimate recombination Recombination with non-homologous regions of DNA. Each of the polynucleotide strands of double-stranded DNA is religated with DNA from different regions of the same or different chromosomes. (*see also* homologous recombination)

image plate A kind of area detector; a phosphor-impregnated plate that can be scanned to quantify its radiation exposure, and can then be regenerated.

α-imino acid A carboxylic acid that bears a substituted amino group on its α-carbon; commonly proline, hydroxyproline, sarcosine.

imino group A secondary amine, e.g. the >NH group of proline.

iminum ion (*see* a-type ion)

immediate-early stage The first stage of response of a cell to viral infection, followed by the *delayed-early stage* and the *late stage*. Each stage is characterized by transcription of specific viral genes.

immobile DNA junction A branched double-stranded DNA structure, a chi-form in which each of four polynucleotide chains is complementary to and base-paired with the 3'-end of one polynucleotide and the 5'-end of another; unlike in a recombination intermediate, the junction cannot migrate (i.e. slide its branch point along the duplex) without breaking many more Watson–Crick base-pair hydrogen bonds as it advances than can re-form behind it. (*see also* Holliday model)

immunoaffinity chromatography A variant of affinity chromatography in which an antibody coupled to the stationary phase adsorbs an antigen which is subsequently eluted at a different pH or a higher salt concentration. Specific antibodies may similarly be prepared on columns of immobilized antigens.

immunoblotting (= western blotting)

immunodiffusion A method for testing the identity of several antigens using an *Ouchterlony plate*, a glass plate covered with a thin layer of agar on which is incised a hole surrounded by several other holes that serve as wells. In the central well is placed an antibody, and the antigens in question are placed in the surrounding wells; the antibody and antigens diffuse towards each other until they meet and form a visible precipitin line. An antigen in an adjacent well forms a precipitin line that is continuous with the first only if the two antigens are identical. Similarly, the plate can be used to test the identity of antibodies against a single antigen in the central well.

immunoelectrophoresis A method for analysis of antibody or antigen mixtures. A solution of antibodies is electrophoresed in a gel, then exposed to an antigen solution placed in a trough cut into the gel parallel to the direction of the separation. The antibodies and antigens diffuse into the gel, and where they meet they precipitate and form a visual or stainable arc.

immunofluorescence The visualization of an antigen in a histological section or on a polyacrylamide slab by allowing it to combine with a fluorophore-tagged antibody. In *indirect immunofluorescence* the antibody is not itself labelled but is the target of another antibody which is labelled, e.g. detection of an antigen by a specific human antibody that depends upon the subsequent attachment to the antigen–(human antibody) complex of a rabbit antibody directed against a broad range of human immunoglobulins.

immunogen (= antigen)

immunoglobulin One of a member of five general classes of serum and blood cell proteins that recognize foreign compounds.

immunoglobulin fold The characteristic supersecondary structure of each of the domains of immunoglobulins that consists of seven β-pleated sheets in two antiparallel layers.

immuno-PCR A very sensitive method for detection of antigens. The antigen is immobilized in wells of a microtitre plate, exposed to its specific monoclonal antibody, then to a chimaeric protein that consists of a protein A moiety, which will bind tightly to the immunoglobulin G antibody, and a streptavidin moiety, which will bind to a biotinylated DNA fragment. The fragment is amplified by PCR, separated by gel electrophoresis and stained with ethidium bromide. The chimaeric protein, biotinylated DNA fragment and PCR primers may be used to assay any antigen–antibody pair. The system is sensitive to as few as 580 antigen molecules (10^{-21} mol).

• Sano, T., Smith, C.L. and Cantor, C.R. (1992) Science **258**, 120–122

immunoprecipitation A purification technique that separates antigenic material from a soluble mixture by precipitation with an appropriate antibody; an essential step in radioimmunoassays.

immunotoxin (chimaeric toxin) A conjugate of an immunoglobulin and a toxin which is designed to mimic a type 2 ribosome-inactivating protein; the immunoglobulin recognizes a target cell and the toxin inactivates it.

• Sweeney, E.B. and Murphy, J.R. (1995) Essays Biochem. **30**, 119–131

imprinting The process by which some autosomal genes are rendered nonequivalent. The paternal or maternal allele is not expressed, or is expressed differently in different tissues. The phenomenon is possibly directed by an *imprinting box*, a parent-specific polynucleotide sequence that instructs an *imprinting factor*, which may act by methylation, and which will reversibly activate or inactivate the gene inherited from only one of the parents.

• Barlow, D.P. (1995) Science **270**, 1611–1613; Reik, W. and Allen, N.D. (1994) Curr. Biol. **4**, 145–147

inborn error An inherited defect in a metabolic enzyme or other gene product.

incision (*see* excision)

included mutation A mutation that is expressed in a protein that is itself a subunit of a larger complex.

inclusion body An amorphous deposit in the cytoplasm of a cell; an aggregated protein appropriate to the cell but damaged, improperly folded or liganded, or a similarly inappropriately processed foreign protein, such as a viral coat protein or recombinant DNA product. (*see also* chaperone machine; Heinz body)

indirect bilirubin (*see* van den Bergh reaction)

indirect calorimetry Evaluation of the heat evolved by a subject by measurement of oxygen consumption, carbon dioxide evolution and excretion of nitrogenous metabolites. (*see also* direct calorimetry)

indirect end labelling A method to locate, between restriction sites, binding sites on DNA, e.g. the positions of nucleosomes on a minichromosome. The DNA protected by binding proteins is cleaved in susceptible, i.e. unprotected, regions by non-specific chemical or enzymic methods, the protecting protein is removed, and a restriction nuclease introduces reference points recognized by complementary polynucleotide probes for subsequent analysis of the sequences from the restriction sites towards the sites at which it was initially cleaved in the unprotected regions.

indirect van den Bergh reaction (*see* van den Bergh reaction)

induced-fit theory (rack model) A model for interaction of a protein with a ligand in which the protein binds preferentially with one of several conformers of the ligand; especially enzyme action in which the substrate is proposed to bind to a conformation of the enzyme that produces a strain on the substrate, focused on the bond that must be broken to effect catalysis; also known as the *handshake model*, and contrasted with the *lock and key theory*, in which both protein and ligand are imagined to be inflexible.

inducer (*see* induction)

inducible enzyme An enzyme whose rate of synthesis can be controlled by an inducer. (*see also* constitutive enzyme; induction)

induction The production of an enzyme in response to the presence of a particular compound, i.e. the *inducer*, or a condition, e.g. heat.

inhibitor In enzymology, a compound, or even a macromolecule, that blocks the action of an enzyme by reversible attachment in such a way as to prevent binding by the substrate (*competitive inhibition*), or by prevention of the reaction even if the substrate can still bind (*non-competitive inhibition*).

initial velocity The rate of an enzymic or chemical reaction as it begins, i.e. with all concentrations of substrates and products at defined levels.

initiation codon A trinucleotide sequence, AUG, that signals the start of translation of a protein; the codon for methionine in eukaryotes and for N-formylmethionine in prokaryotes.

initiation complex The ribosome charged with mRNA, initiation factors and methionyl-tRNA (in eukaryotes) or N-formylmethionyl-tRNA (in prokaryotes, mitochondria and chloroplasts) that is bound to the initiation codon. (*see also* pre-initiation complex)

initiation factor An accessory protein that is necessary for assembly of the ribosome–mRNA complex and the start of protein synthesis.

initiator A site, upstream from a structural gene, for attachment of a protein that stimulates initiation of transcription.

initiator tRNA The tRNA that recognizes the AUG start codon. In prokaryotes it can be charged to bear a formylmethionine residue (methionine in eukaryotes) which will become the N-terminus of the new protein.

inner filter effect A fluorescence spectroscopy phenomenon; the decrease in fluorescence emission seen in concentrated solutions due to the absorption of exciting light by the fluorophore that is close to the incident beam and which significantly diminishes light that reaches the sample further away from it.

insertion mutation A mutation caused by the presence of one or several extra nucleotide residues in the DNA sequence, which may result in a frameshift.

insertion sequence A small transposon, bordered at its 3'- and 5'-ends by inverted terminal repeats.

in situ 'In place'; descriptive of, for example, perfusion of the liver within the abdominal cavity rather than excision of the organ and perfusion *in vitro*.

in situ **PCR** The performance of PCR on fixed preparations on a microscope slide.
• Taylor, G.R. and Logan, W.P. (1995) Curr. Opin. Biotechnol. **6**, 24–29

instructive theory A now discarded hypothesis that explains antibody specificity by the folding of an antibody structure, which has the potential for many specificities, to the unique surface of a particular antigen. (*see also* selective theory)

intasome A nucleoprotein complex active in integration of bacteriophage DNA into host DNA. Negatively supercoiled host DNA is forced to cross back over itself at points called *nodes*, which are sites for binding of proteins that facilitate the recombination with phage DNA or result in crossing-over of the host strands and formation of a knotted DNA.
• Nash, H.A. (1990) Trends Biochem. Sci. **15**, 222–227

integral protein (intrinsic protein) A protein intimately embedded in a membrane, frequently even passing through it and emerging on both sides. (*see also* peripheral protein)

intein The protein counterpart of the RNA intron; a polypeptide sequence that is excised by a self-catalysing mechanism from the primary translation product. (*see also* protein splicing)
• Hickey, D.A. (1994) Trends Genet. **10**, 147–149; Cooper, A.A. and Stevens, T.H. (1995) Trends Biochem. Sci. **20**, 351–356

intercalating agent A compound that acts by inserting itself between adjacent bases in a DNA chain; during replication it is capable of causing either a deletion or an insertion mutation.

interchromomere (*see* AT queue)

interferon One of a group of small antiviral proteins synthesized and secreted by cells following viral infection.

intermediary metabolism The individual enzymic reactions that in a tissue or inside a cell transform one metabolite into another, e.g. the conversion of lactate into glucose by the liver, the conversion of leucine into acetoacetate by muscle, the conversion of sucrose into ethanol by yeast.

intermediate filament A structure intermediate in diameter between a thin filament and a microtubule, usually 7–11 nm in diameter; a neurofilament, keratin molecule, etc.

internal control region A regulatory site of a gene that interrupts a structural gene.

internal guide sequence A polynucleotide sequence near the 5'-end of group I introns that pairs with sequences of the upstream exon in an intermediate of the self-splicing process. (*see also* self-splicing)

internal signal sequence An internal polypeptide sequence of a protein that is responsible for its targeting to the appropriate locus within a cell after synthesis. (*see also* signal sequence)

interphase nucleus mapping A variant of the usual fluorescence *in situ* hybridization technique in which non-mitosing cells are used to visualize a genetic marker on a (non-visible) chromosome by use of a fluorescence-labelled polynucleotide probe that hybridizes to and indicates the locus of a gene on a chromosome.
- Liehr, T., Grebl, H. and Rautenstrauss, B. (1995) Trends Genet. **11**, 377–378

inter-resource duplex (*see* subtractive DNA cloning)

interspersed repetitive sequence-PCR (IRS-PCR) A method for amplification of human DNA sequences in hybrid somatic cells, using primers directed at species-specific repetitive sequences: a short interspersed repeat element'(e.g.

Alu) sequence, a long interspersed repeat sequence or a combination of both.
- Nelson, D.L. (1991) Methods Companion Methods Enzymol. **2**, 60–74

intervening sequence region (= intron)

intimacy model (*see* altered-self hypothesis)

intrinsic factor A glycoprotein secreted by the gastric mucosa that assists in the absorption of vitamin B_{12} (extrinsic factor) in the intestine; absent from subjects with pernicious anaemia.

intrinsic pathway The blood clotting cascade that is initiated by a factor that can be generated within the blood itself. (*see also* extrinsic pathway)

intrinsic protein (= integral protein)

intron From *intervening sequence region*; originally defined as a non-coding polynucleotide sequence that interrupts the coding sequences, the *exons*, of a gene. More recent evidence indicates that, in some cases, introns may also be translated, so a more cautious definition of an intron is merely a DNA sequence whose transcribed sequence is physically separated from that of exons during RNA processing. (*see also* spliceosome)
- Moore, M.J. (1996) Nature (London) **379**, 402–403

intron-early (*see* shuffling)

intron editing A change in the reading of the sequence coded by DNA directed by an adjacent intron. The intron, which has a sequence complementary to part of the exon, forms a double-stranded stem structure that is subject to enzymic alteration of a coding base, e.g. deamination of an adenine, before the intron is eliminated by splicing. (*see also* spliceosome)

intron homing The direction of introns, as

mobile genetic elements, to sites in DNA. The process is dependent upon proteins that are encoded by open reading frames of the introns: endonucleases in the case of group I introns and reverse transcriptases in the case of group II introns. (*see also* protein splicing)

- Belfort, M. and Perlman, P.S. (1995) J. Biol. Chem. **270**, 30237–30240; Pyle, A.M. (1996) Nature (London) **381**, 280–281

intron-late (*see* shuffling)

inverse PCR A PCR protocol for amplification of sequences that flank a single known sequence. The source DNA is digested with a restriction endonuclease, circularized and ligated. Primers are constructed that will hybridize to the ends of the known sequence, but will be extended in opposite directions into the flanking sequences and around the circle. Cycles of polymerization, denaturation and rehybridization amplify the sequence between the restriction sites. (*see also* ligand-mediated PCR; vectorette PCR)

- Ochman, H., Medhora, M.M., Garza, D. and Hartl, D.L. (1990) in PCR Protocols: a Guide to Methods and Applications (Innis, M.A., Gelfand, D.H., Sninsky, J.J. and White, T.J., eds.), pp. 219–227, Academic Press, San Diego

invertasome A nucleoprotein complex that mediates inversion of a DNA sequence, bracketed by two recombination sites, within the genome of an organism.

- Heichman, K.A. and Johnson, R.C. (1990) Science **249**, 511–517

inverted terminal repeat A polynucleotide sequence that is repeated in reverse sequence; the flanking sequence of a transposon.

in vitro 'In glass'; taken into an experimental situation, as opposed to *in situ* or *in vivo*.

***in vitro* selection** (= systematic evolution of ligands by exponential enrichment; *see* cyclic amplification and selection of targets)

in vivo In the living organism, as opposed to *in vitro*. In biochemistry the term often refers to activity within a living cell.

iodine number A measure of the unsaturation of a lipid; the number of grams of iodine that react with 100 g of an olefin.

iodine test The reaction of some polysaccharides, e.g. starch or glycogen, with iodine to give a red to purple colour; diagnostic of an α-linked polysaccharide.

ion-exchange chromatography A technique for separation of charged compounds or macromolecules in solution according to their affinities for a positively charged stationary phase (an *anion exchanger*) or a negatively charged stationary phase (a *cation exchanger*).

ionophore An agent that allows passage of an ion through an otherwise impermeable membrane, e.g. the protonophore thermogenin, which makes the inner membrane of brown adipose tissue mitochondria permeable to protons.

iontophoresis The movement of an ion driven by an electrical potential.

ion trap A mass spectrograph in which vaporized ions are confined by an AC electric field. As the radio-frequency voltage that confines the ion is changed, the ion will escape the trap at a voltage characteristic of its mass-to-charge ratio, and be detected. Sequences of polypeptides and polynucleotides may be determined by expulsion from the trap of all but one kind of large mass ion, fragmentation of this ion by collision with helium atoms and analysis of the mass spectrum of the resulting fragments. A variant of this electrical-type ion trap uses a magnetic field to confine ions. (see also tandem mass spectrometry)

- Cooks, R.G., Glish, G.R., McLuckey, S.A. and Kaiser, R.E. (1991) Chem. Eng. News **69 (12)**, 26–41

IRD Inter-resource duplex. (see subtractive DNA cloning)

iron–sulphur cluster The complex of iron and sulphur atoms in an iron–sulphur protein.

iron–sulphur protein One of a class of proteins in which one to four iron atoms, depending upon the protein, are co-ordinated with the sulphur atoms of cysteine and/or with sulphide ions; e.g. succinate dehydrogenase, aconitase.

IRS-PCR (= interspersed repetitive sequence PCR)

islet A histological structure; usually the islet of Langerhans in the pancreas which consists of glucagon-secreting α-cells and insulin-secreting β-cells.

isoacceptor One of two or more tRNAs that accept the same amino acid.

- Saks, M.E., Sampson, J.R. and Abelson, J.N. (1994) Science **263**, 191–197

isobol (see isobologram)

isobologram A representation of enzyme kinetic data as a plot of varying substrate and inhibitor concentrations that give constant activity.

isocratic At constant concentration.

isoelectric focusing (IEF) A technique to separate proteins by their isoelectric points. Proteins are electrophoresed through a gel that has imposed upon it a pH gradient stabilized by the inclusion of polyampholytes, which themselves have a range of isoelectric points. Each protein migrates until it reaches the pH region of its isoelectric point.

isoelectric point The pH value at which an amphoteric compound is electrically neutral.

isoenzyme (isozyme) One of two or more structural variants of an enzyme that can occur in the same individual, often relatively specific for a particular tissue, each with unique kinetic characteristics; e.g. the heart (H)- and muscle (M)-type lactate dehydrogenase subunits.

isoform A tissue-specific form of a non-enzymic protein; analogous to an isoenzyme.

isohydric shift The transformations in the erythrocyte that accommodate the uptake and partial binding of carbon dioxide while maintaining a relatively constant pH.

isomerase One of a class of enzymes that rearrange the bonds of their substrates, e.g. an epimerase.

isomorphous replacement An X-ray crystallographic technique to assist in solving structures from a diffraction pattern. One or more heavy-metal atoms are inserted in the crystal (replacement) without disturbing the arrangement of the other atoms of the structure (isomorphous). In single isomorphous replacement a single derivative is used to solve a molecular structure; in multiple isomorphous replacement several derivatives are used. (see also difference Patterson map)

iso-osmotic (see osmolarity)

isopeptide bond An amide bond between amino acids that employs a non-α-amino or a non-α-carboxy group, e.g. the γ-carboxy–ε-amino cross-link of hard fibrin clots.

isoprene rule The observation, first based on inspection of structures and later on experimentation, that many biological substances (e.g. sterols, tocopherols, rubber) appear to have been assembled from multiple isoprene units. (*see also* terpene)

isoprene unit The branched five-carbon chain that is a recognizable component of terpenes and other compounds derived from isopentenylpyrophosphate, the biosynthetic isoprene unit. (*see also* terpene)

isopycnic centrifugation (= density-gradient centrifugation)

isoreceptor One of two or more homologous plasma membrane or cytosolic receptors that have the same or altered functions; analogous to an isoenzyme.

isosbestic point In spectroscopy, the wavelength at which two or more components of a solution have the same molar absorption coefficient.

isoschizomers A pair of restriction endonucleases that recognize the same palindromic sequence, although in some cases one member of a pair may differ from the other in its response to methylation of the recognition sequence, or may cleave a different phosphodiester bond within the target sequences.

isotachophoresis (*see* electrophoresis)

isotherm A binding curve; at constant temperature, the concentration-dependence of the binding of one material to another, e.g. a gas at various pressures to a solid surface, an amino acid at various concentrations of a neutral salt to a chromatography matrix.

isotonic (*see* osmolarity)

isotope A variant of an atom, chemically identical but with a different atomic mass; often radioactive.

isotope dilution assay A method to determine the amount of a substance in a mixture by addition to it of a known amount of the same material, isotopically labelled and of known specific radioactivity. The material is subsequently isolated in pure form and its new specific radioactivity is determined. From the degree of dilution and the amount of labelled material added, one may calculate the original concentration.

isotope effect The effect on kinetics of the substitution of one isotope for another in a bond that must be broken in a rate-limiting step (*primary isotope effect*); especially prominent in the case of cleavage of a bond to a deuterium atom compared with a hydrogen atom. Smaller effects are seen in the case of a substitution of an isotope for a non-reacting atom (*secondary isotope effect*); effects are also seen on parameters other than the rate-limiting step. (*see also* solvent isotope effect)

isotropic Descriptive of a physical property that is independent of the angle of observation. (*see also* anisotropic)

isotype An immunoglobulin molecule classified according to the amino acid sequences of the constant regions i.e. α, β, δ, ε, γ and μ heavy chains and κ and λ light chains. (*see also* allotype; idiotype)

isozyme (= isoenzyme)

IVS Intervening sequence. (= intron)

J

J-chain A polypeptide found in the dimeric immunoglobulin A and the pentameric immunoglobulin M, involved in joining together the subunits of each multimer.

jigsaw-puzzle model (*see* framework model)

jumping gene (= transposon)

jumping library A tool for the mapping and sequencing of large stretches of a chromosome. Starting at one point in the polynucleotide sequence, a distal point is located without having to first sequence the intervening bases. The double-stranded DNA is divided into large fragments with a rare cutter and one end of each piece is ligated to a selective marker and cyclized, then cut with a frequent cutter. The marker allows isolation of the two ends of the original large piece without the intervening sequence.

jumping PCR A technique for reconstruction and replication of randomly damaged double-stranded DNA, e.g. that found in mummified tissues or museum samples. In the first PCR cycle each primer is extended until a break in the template is encountered, but in the second cycle the newly synthesized strands find other templates that are intact in the region where the first templates were damaged. Each cycle extends the length of newly synthesized strands until complete strands are formed and PCR can begin its exponential synthesis.

junk DNA DNA for which a function has yet to be identified. An elaborate nomenclature has been proposed for this DNA: any stretch of DNA is a *nuon*; DNA with an evolutionary potential for development of function is a *potonuon*; DNA that was previously non-functional, or that had a completely different function and has been evolutionarily co-opted for a new function (a process termed *exaptation*), is a *xaptonuon*; and a non-co-opted potonuon is a *naptonuon*.

• Brosius, J. and Gould, S.J. (1992) Proc. Natl. Acad. Sci. U.S.A. **89**, 10706–10710; Nowak, R. (1994) Science **263**, 608–610; Flam, F.F. (1994) Science **266**, 1320

K

K_a (= association constant)

kallikrein (*see* kinin; kininogen)

karyokinesis Division of the nucleus of a cell during mitosis.

karyophile A substance that can diffuse into and be bound in the nucleus of a cell, presumably due to its affinity for a non-diffusible nuclear compound.

karyotype The display of chromosomes, arranged from largest to smallest, obtained by cutting out and arranging photomicrographs of chromosomes.

katal A measure of enzyme activity; the conversion of 1 mol of substrate per s under specified conditions. (*see also* unit)

kbp (kb) Kilobase pairs (kilobases).

k_{cat} (= catalytic rate constant)

k_{cat} **inhibitor** (*see* mechanism-based inhibitor)

K_d (= dissociation constant)

kDa Kilodalton.

ketal (*see* hemiketal)

ketimine form (*see* quinimine form)

ketogenic amino acid An amino acid whose carbon skeleton can be converted, at least in part, into ketone bodies. (*see also* glucogenic amino acid)

17-ketogenic steroid A steroid that on oxidation with sodium bismuthate will form a 17-keto (-oxo)-steroid; usually a 17-hydroxy-, 20-oxo- or 17,20-dihydroxy-steroid.

ketone body Acetoacetate, β-hydroxybutyrate or acetone.

ketosis An accumulation in the blood of ketone bodies, indicative of a metabolic dysfunction such as uncontrolled diabetes mellitus or starvation.

killer T-cell (*see* T-cell)

kilobase (kb) A measure of DNA chain length; 1000 nucleotide residues.

kinase An enzyme that uses ATP to phosphorylate a substrate; also, in older literature, an enzyme that activates its substrate, e.g. enterokinase. (*see also* protein kinase C)

kinesis (*see* protein kinesis)

kinetic assay An enzyme-based assay that measures the amount of substrate present by correlation of the rate of reaction with the known dependence of the rate on substrate concentration, usually under first-order conditions. (*see also* end-point assay)

kinetic enrichment (*see* subtractive DNA cloning)

kinetic partitioning The hypothetical basis of recognition of proteins by molecular chaperones. Soon after synthesis, a cytosolic protein achieves its native conformation and thus avoids binding to the chaperone, while other proteins, e.g. those with leader sequences, fold into native structures much more slowly, which affords the chaperone the opportunity to bind them.

• Hardy, S.J.S. and Randall, L.L. (1991) Science **251**, 439–443

kinetic proofreading A mechanism for assuring fidelity in directed macromolecular synthesis; e.g. in protein synthesis a well matched anti-codon–codon pair, as opposed to a poorly matched pair, will remain associated until commitment to the next step of synthesis, which is timed by an internal 'clock', the hydrolysis of bound GTP. (*see also* G-protein)

- Bourne, H.R., Sanders, D.A. and McCormick, F. (1990) Nature (London) **348**, 125–132

kinetics The study of rates of reactions.

kinin A hypotensive plasma peptide that dilates small blood vessels and increases capillary permeability, e.g. bradykinin; formed from its precursor, a *kininogen*, by the proteolytic action of a *kallikrein*.

kininogen A protein that contains the sequence Arg-Pro-Pro-Gly-Phe-Ser-Pro-Arg which has inhibitory activity against cysteine proteinases. A kininogen is a precursor of a *kinin*, which is a decapeptide or undecapeptide produced by the proteolytic activity on a kininogen by a *kallikrein*. An *L-kininogen* is a low-molecular-mass kininogen and an *H-kininogen* is a high-molecular-mass kininogen, e.g. T-kininogen is one of the H-kininogens in the rat. (*see also* alpha$_1$-cysteine proteinase inhibitor; cystatin; thiostatin)

kissing complex The interaction of two RNA molecules by hybridization of complementary sequences from a loop of each.

Klenow fragment The product of limited proteolysis of *E. coli* DNA polymerase I that retains the polymerase and 3'-to-5' exonuclease (proofreading) activity but is missing the 5'-to-3' nuclease activity.

knock-out An animal or plant from which a certain gene has been deliberately deleted, in order to observe, as a test for the gene's physiological significance, the viability or phenotype of the organism.

K_m (= Michaelis constant)

KNF model (= Koshland model)

knot (*see* node)

Koshland model (KNF model) A sequential model for the basis of co-operativity in multimeric binding proteins, originally presented as a description of oxygen binding to each of the haem groups of the haemoglobin tetramer. Unlike in the earlier Monod–Wyman–Changeux concerted model, each monomer may assume a low-affinity or a high-affinity conformation. A high-affinity ligand-bound monomer favours the high-affinity conformation of adjacent unbound monomers. In the absence of a ligand (oxygen in the case of haemoglobin) the subunits exist in a low-affinity 'taut' conformation. The ligand may bind to as many as two monomers in this conformation, at which point all four monomers convert from the low-affinity form to the high-affinity 'relaxed' conformation.

Krebs cycle (= tricarboxylic acid cycle)

Krebs–Kornberg cycle (= glyoxylate cycle)

Krebs urea cycle (= urea cycle)

kringle A protein folding motif, as visualized in two dimensions, of loops formed by multiply disulphide-bridged sequences of several circulating proteins, e.g. α_2-macroglobulin, complement component C3, prothrombin, tissue-type plasminogen activator.

Kunkel mutagenesis (dUTP system mutagenesis) A method to construct a site-specific mutated gene and to select it away from the wild-type. A strain of *E. coli* with only weak dUTPase activity has high levels of dUTP and consequently incorporates dUTP into a plasmid DNA in competition with dTTP. This plasmid is used in an *in vitro* system as a template for DNA synthesis with a primer into which the desired mutation has been incorporated. The duplex, containing the wild-type sequence with Us replacing Ts (*U-DNA*) complementary to a strand with the usual DNA bases and incorporating the mutation, is introduced into a wild-type bacterium that can remove the inappropriate bases from, and thus inactivate, the U-DNA strand and leave the mutated strand to be replicated.

L

L- (*see* D-)

label A chemical modification or isotope substitution that makes possible the identification of a molecule, moiety or atom throughout its metabolic transformations or transport.

lability Instability to heat, shear or other physical or chemical stress.

***lac* operon** The structural and associated regulatory genes that control the ability of *E. coli* to live on lactose; a subject of classical experimentation.

ladder The pattern of bands seen on a polynucleotide sequencing gel that differ by integral numbers of nucleotides. (*see also* Maxam–Gilbert method; Sanger method)

Laemmli gel electrophoresis (= sodium dodecyl sulphate/polyacrylamide-gel electrophoresis)

***laevo* (*levo*)-rotation** (*see* optical rotation)

lagging strand The discontinuous DNA strand that is synthesized at a fork during replication, the direction of synthesis of which is opposite to that of the movement of the fork. (*see also* leading strand; Okazaki fragment; semi-discontinuous)

LAGLI-DADG sequence A conserved sequence in yeast mitochondrial unassigned reading frames, about 1 kb long, that is part of an intron the 5′-end of which has the consensus sequence coding for the LAGLI pentapeptide and the 3′-end that for a DADG tetrapeptide.

lambda phage (*see* bacteriophage)

lamellipodium (ruffled edge) A fin-like projection of the plasma membrane of a eukaryotic cell.

lampbrush chromosome A chromosome in which the chromatin extends out from a central axis into many loops, where transcription and other activity occurs.

lap-joint The non-covalent association of sticky-ended DNA duplexes.

lariat (*see* self-splicing)

laser-assisted desorption mass spectrometry (*see* mass spectrometry)

late stage (*see* immediate-early stage)

latent Hidden or cryptic, e.g. the enzymic properties that are expressed *in vitro* only when the vesicle that encapsulates the enzymes is lysed.

lathyrism A skeletal malformation due to dietary β-aminopropionitrile (present in the seeds of sweet peas, *Lathyrus odoratus*) which interferes with the proper cross-linking of collagen.

Laue crystallography (time-resolved crystallography) The use of X-rays from a synchrotron to collect diffraction data in very short times, sometimes measured in milliseconds, for time-resolved crystallography of proteins in order to observe internal motion in the protein associated with conformational changes and/or catalysis.

- Gruner, S.M. (1994) Curr. Opin. Struct. Biol. **4**, 765–769

lawn The thick coverage of an agar (or other gel medium) plate with bacteria; usually as a test system for a lysogenic bacteriophage. (*see also* plaque-forming unit)

layer line A feature of an X-ray diffraction pattern of a fibre. For fibres that contain aligned helices, the reflections appear as a pattern of spots arrayed in lines perpendicular to the fibre axis and the separation of the lines is proportional to the rise of the helix.

LCR (= ligase chain reaction)

LD$_{50}$ The dosage of a toxic material that is lethal to 50% of the animals, or cells, studied.

LD-MS Laser-assisted desorption mass spectrometry. (*see* mass spectrometry)

leader sequence (= signal sequence)

leading strand The continuous DNA strand synthesized at a fork during DNA replication. (*see also* lagging strand; semi-discontinuous)

leaflet One half of a lipid bilayer.

lectin A protein, other than one of the immunoglobulins, that binds to the non-reducing ends of a specific oligosaccharide, e.g. that of a glycoprotein or cell surface. Lectins are usually of plant origin, although some have been identified in animal tissues and bacteria.

• Sharon, N. and Lis, H. (1995) Essays Biochem. **30**, 59–75

left-handed helix (*see* right-handed helix)

leghaemoglobin A single-chain oxygen-binding haem protein of legumes.

Leloir pathway The metabolic route by which galactosyl and glucosyl units are interconverted; the transformation of galactose and ATP to glucose-1-phosphate via the galactokinase, galactose-1-phosphate uridyltransferase and UDP-galactose-4-epimerase reactions.

• Frey, P.A. (1996) FASEB J. **10**, 461–470

Lepore haemoglobin A class of haemoglobin in which there has been a faulty crossing-over between the adjacent genes for β- and δ-globins, producing a globin with the N-terminus of δ-globin and the C-terminus of β-globin. *Anti-Lepore haemoglobins* are mutants with the N-terminus of β-globin and the C-terminus of δ-globin.

leucine scissors A supersecondary structure of some DNA-binding proteins; a coiled coil of two parallel α-helices, each of about 30 residues in length and containing leucine residues at every seventh position. The side chains interact with non-polar side chains of the other helix to hold the two together by hydrophobic forces.

leucine zipper (= leucine scissors)

leukotriene One of a series of arachidonic acid metabolites; contains three conjugated double bonds and has been oxygenated through the action of a lipoxygenase.

levo-rotation (= *laevo*-rotation; *see* optical rotation)

library (*see* cDNA library; combinatorial; genomic library; shape library; small molecule library)

licencing The limiting of replication of the eukaryotic genome to once per cell cycle. The proteins that impose this effect are termed *replication licencing factors* (*RLFs*), but their mode of action is unknown.

LIC-PCR (= ligation-independent cloning of PCR products)

ligand A small molecule or ion that binds to a protein or other structure.

ligand-mediated PCR (LMPCR) A method for amplification of a DNA fragment or a series of fragments with a single known primer-binding site. The fragment is denatured and hybridized with the primer. Extension of the primer with a polymerase that leaves a blunt end allows subsequent ligation to a *linker-primer*, which is a duplex oligodeoxynucleotide that will become the attachment site of a second primer in subsequent PCR cycles. (*see also* anchored PCR; inverse PCR)

ligase One of a class of enzymes that join two substrate molecules in energy- (usually ATP-) dependent reaction, e.g. an amino acyl-tRNA synthetase, a carboxylase; in molecular biology, an enzyme that attaches the 3'-end of one polynucleotide to the 5'-end of another. (*see also* synthetase)

ligase chain reaction (LCR) A technique for detection and amplification of target DNA sequences. Two oligodeoxynucleotides are synthesized which between them are complementary to the entire target sequence, one to the 5'-side and one to the 3'-side. If the target sequence is present in the DNA sample under examination, the oligonucleotides will bind to it at moderate temperature with their ends abutting in the centre, and a heat-stable ligase will join them into a complete polynucleotide. No ligation will occur if the target sequence is absent or if the match between synthetic oligonucleotides and target sequence is imperfect in the region where they abut. At an elevated temperature the new polynucleotide dissociates from the original DNA template, and upon cooling to moderate temperature it and the original DNA serve as templates for a second cycle of hybridization, ligation and thermal dissociation. At each cycle there is a doubling of the number of new complete polynucleotides, so after 20 cycles there can be a 10^6-fold amplification. The half-sequences may be constructed with markers on the non-ligating ends to assist isolation (e.g. biotin) or detection (e.g. a fluorophore). (*see also* polymerase chain reaction)

• Weiss, R. (1991) Science **254**, 1292–1293

ligation-independent cloning of PCR products (LIC-PCR) A method for cloning complex mixtures of PCR products to achieve a library of recombinant clones. PCR-amplified sequences, e.g. those bracketed by *Alu* sequences, are generated with primers that include sequences that contain a relatively long (e.g. 12 nucleotides) sequence devoid of G residues; similarly, a vector is cleaved at a restriction site and PCR-amplified with two different primers (different to prevent premature cyclization) that lack C residues in their 3'-ends. The potential inserts and the vector are trimmed with T_4 DNA 3'-5' exonuclease in the presence of dGTP and dCTP respectively to generate overhanging remnants of the primers, which were designed to permit insertion of the PCR fragments into the vector. As the overhangs are relatively long and quite stable, ligation before transformation is unnecessary.

• Aslanidis, C. and de Jong, P.J. (1990) Nucleic Acids Res. **18**, 6069–6074

ligation-mediated PCR A PCR technique, similar to anchored PCR, applied to double-stranded DNA molecules for which the sequence of only one end is known, e.g. in *in vivo* footprint analysis to detect sites protected by a DNA-binding protein from dimethyl sulphate reaction and thus subject to subsequent chemical cleavage; a specific oligonucleotide is ligated to the unknown end and amplification is accomplished by the PCR procedure using primers complementary to the known end and the newly ligated oligonucleotide.

• Mueller, P.R. and Wold, B. (1991) Methods Companion Methods Enzymol. **2**, 20–31

light chain (*see* heavy chain)

light-chain library (*see* repertoire cloning)

light reaction (= Hill reaction)

light scattering A phenomenon of solutions of macromolecules; used analytically to determine the molecular mass and shape of a macromolecule from the effect on the intensity of scattered light of the wavelength of light and the angle of its scatter. In *Rayleigh scattering* the interaction of a photon with the molecule is elastic, i.e. there is no energy transfer from photon to molecule. In *Raman scattering* the interaction is inelastic, i.e. there is energy exchange, usually to vibrational energy (infrared) of the molecule, and the scattered light is then of longer wavelength.

light subunit The smaller of the two ribonucleoprotein complexes that make up a ribosome; more generally, the smallest of the subunits of any complex. (*see also* heavy subunit)

lignin A random polymer in woody plants formed by oxidation of shikimic acid metabolites to form carbon–carbon and carbon–oxygen cross-links.

limit dextrin The partially digested polysaccharide that is left after phosphorylase removes the outer tiers of glycogen or amylopectin and exposes the $\alpha 1 \rightarrow 6$ branch points.

limiting polarization The polarization of a fluorophore when it is immobile. (*see also* Perrin plot)

LINE (= long interspersed repeat element)

Lineweaver–Burk plot A linearized display of kinetic data showing dependence of enzyme activity on substrate concentration, i.e. reciprocal of rate against reciprocal of substrate concentration, therefore also termed a *double-reciprocal plot*. (*see also* Eadie–Hofstee plot)

line width The frequency range over which energy absorbance occurs, or over which a spectroscopic peak is observed; frequently measured at half the peak height.

linkage In genetics, the more-frequent-than-random occurrence of two traits together due to their proximity on the same chromosome. The likelihood of a recombination event separating the two traits decreases with their increasing proximity on the chromosome. (*see also* linkage disequilibrium)

linkage disequilibrium In genetics, the more-frequent-than-random occurrence of two traits together due to selective pressure or to a close linkage of one of the genes to a third gene for which there is selection. (*see also* linkage)

linkage group A collection of genes that are inferred to be located together on a single chromosome because of the pattern of their inheritance.

linkage map A description of a genome in which many phenotypes are screened to determine the frequency with which genes occur together; these can then be placed in appropriate proximity to one another.

• Weissbach, J., Gyapay, G., Dib, C., et al. (1992) Nature (London) **359**, 794–801

linker-primer (*see* ligand-mediated PCR)

linker-scanner mutation A method to assess the importance of various regions of a protein to its function by introduction of a dipeptide sequence into random sites of the protein. A plasmid that contains a structural gene is partially cleaved by a restriction endonuclease to achieve a single cleavage per coding region. The double-stranded DNA linkers, which are two-anticodon sequences between restriction nuclease-specific ends, are then patched into the gaps left by the cleavages.

linking clone A tool for the localization of restriction fragments within a physical map of the genome of an organism. A cloned fragment that contains very few of an already rare restriction site is used as a hybridization probe for chromosomal DNA digested with the rare cutter.

• Billings, P.R., Smith, C.L. and Cantor, C.R. (1991) FASEB J. **5**, 28–34

linking number (*L*) A topological property of double-stranded DNA; the number of right-handed turns one DNA strand makes around the other. For relaxed DNA the number is equivalent to what it would be for linear, unstressed, DNA, i.e. 1 per 10 base pairs. L is equal to the twisting number plus the writhing number ($L = T + W$).

link protein A protein that holds a subunit of a proteoglycan to the central hyaluronic acid molecule. (*see also* core protein)

lipid A natural substance that is poorly soluble in water but is soluble in organic solvents; lipids include fatty acids, triacylglycerols, phospholipids, waxes and some hormones and vitamins.

lipid bilayer (bimolecular sheet) A synthetic or natural membrane in which amphipathic lipids are arranged in two layers with their non-polar chains directed inwards, towards each other, and their polar groups directed outwards, towards the aqueous phase. (*see also* fluid mosaic model)

lipid peroxidation The oxidation of polyunsaturated fatty acid side chains, initiated by a free radical such as the hydroxyl radical (OH•), to form a relatively stable carbon free radical, which reacts further with molecular oxygen to form a peroxy radical, and further yet with another fatty acid side chain to generate another carbon free radical. Such peroxidation causes rancidity in foods, damages membranes and, when it takes place in plasma low-density lipoprotein, can lead to atherosclerosis. (*see also* oxidative stress)

lipocalin One of a family of homologous proteins that bind lipophilic substances, including retinol-binding protein, β-lactoglobulin, orosomucoid.

lipoma A cancer of adipose tissue.

lipopolysaccharide (LPS) A constituent of the outer membrane of Gram-negative bacteria; composed of an outward-directed and highly variable oligosaccharide (the *O-antigen*), which is responsible for the antigenicity of the product, linked

to a relatively invariable core oligosaccharide, in turn linked to a less polar
moiety (*lipid A*) that is embedded in the
membrane and is responsible for the
endotoxicity and pyrogenicity of the
product. *R strains* produce colonies with
a rough surface and have truncated
lipopolysaccharides; *S strains* produce
colonies with a smooth appearance.

lipoprotein A complex of lipids and
apolipoproteins that is a transport form of
lipids in blood. Lipoproteins are characterized by their density, which is determined by the lipid portion, and include
high-, low- and very-low-density lipoproteins.

liposome A synthetic micelle formed by
dispersion of a polar lipid in aqueous
solution in such a way that it forms into
lipid-bilayer-encapsulated vesicles; useful for packaging of drugs or enzymes for
introduction into cells.

L-kininogen (*see* kininogen)

LMPCR (= ligand-mediated PCR)

location measure An indication of the position of an amino acid residue within a
proposed folding of a globular protein;
the number of α-carbons within a 14 Å
(1.4 nm) radius, which varies from about
10 for a residue on the surface of a protein to about 80 for a residue well inside
a large globular protein. (*see also* contact
number)

lock and key model A model for enzyme
action that explains the basis of specificity as the exact fit of substrate to a site
on the enzyme that is complementary in
shape and electronic charge. (*see also*
induced-fit theory)

locus The position on a chromosome of a
particular allele.

LOH (= loss of heterozygosity)

London dispersion force The weak interaction of neutral atoms due to the transient
asymmetrical charge distribution within
each atom. (*see also* van der Waals bond)

long interspersed repeat element (LINE) A
recurring species-specific polynucleotide
sequence in the human genome, of
approximately 6.4 kb, present at 10^4 to
10^5 copies. (*see also* small interspersed
repeat element)

long-chain fatty acid (*see* medium-chain
fatty acid)

long terminal repeat (LTR) A polynucleotide sequence found at each end
of an integrated retrovirus genome that
contains the signals for expression of the
viral genome.

loop A packing structure of eukaryotic
DNA that may be identical to a replicon.
It is seen when the DNA is unfolded and
visualized by electron microscopy, and
shows the polynucleotide extending
from closely spaced points of attachment
to the nuclear matrix, which are presumed to be the terminators of replication. Also, in protein chemistry, a short
polypeptide sequence of a protein that
connects one region of secondary structure (α-helix or β-sheet) to another.

loss of heterozygosity (LOH) A test for
chromosomal changes in tumour DNA
and the theory that some tumours arise
from inactivation of both alleles for an
anti-oncogene. Heterozygotes, who
already have one mutation, are most vul-
nerable. The tumour (e.g. breast cancer)
genome, (presumably homozygous) is

compared with the constitutional (e.g. leucocyte) genome (presumably heterozygous) for changes that can be interpreted as a change from the heterozygous to a homozygous state. With an appropriate probe for a site near a suspected mutation, restriction fragment length polymorphism analysis is used to detect the change.

low-angle scattering A technique for determination of the size of particles in solution. The intensity of scattered X-rays from irradiation of a particle, e.g. a ribosome or a virus, is related to its size. In a variant technique, the solution of particles is irradiated with neutrons, the scattering of which varies with the mass of an atom's nucleus and thus can indicate the size of a phosphorus (i.e. nucleic acid)-containing particle or a ^{2}H-labelled protein.

low-barrier hydrogen bond (*see* hydrogen bond)

Lowry protein assay A refinement of the colorimetric method for protein determination that uses the Folin–Ciocalteu phosphomolybdotungstate reagent.

- Lowry, O.H., Rosebrough, N.J., Farr, A.L. and Randall, R.J. (1951) J. Biol. Chem. **193**, 265–275

LPS (= lipopolysaccharide)

LTR (= long terminal repeat)

luciferin A potentially chemiluminescent substrate; the enzymic reaction of an enzyme (a *luciferase*) converts it to an excited state that decays with emission of visible radiation; e.g. the benzothiazole of fireflies which upon adenylation is decarboxylated and deadenylated with the emission of yellow light. The energy of photon emission derives from an oxidation with molecular oxygen that may be the enzymic or a non-enzymic reaction of the product of the enzymic reaction. There is no structural homology within the luciferins or luciferases, which have a wide biological distribution.

luminal (= apical)

luminescence Emission of a photon, of the same or lower energy than the energy that excited it, by an excited state of a chemical compound. (*see also* fluorescence; phosphorescence)

lyase One of a class of enzymes that add one substrate across the double bond of another substrate, e.g. a decarboxylase, a dehydratase, an aldolase. (*see also* synthase)

lymphokine A protein secreted by T-cells that functions in cell-mediated immunity to activate or inhibit leucocytes.

lymphoma A cancer of lymphatic tissue.

lyotropic series (Hofmeister series) The arrangement of anions and cations according to their ability to modify the properties of other solutes such as proteins, probably via their effectiveness in altering the internal structure of the bulk water phase. *Chaotropes* are agents that disrupt water structure, and *cosmotropes* are agents that stabilize it.

- Leberman, R. (1991) FEBS Lett. **284**, 293–294; Parsegian, V.A. (1995) Nature (London) **378**, 335–336

lysate The product of lysis; a broken-cell preparation.

lysis The rupture of the membrane of a cell or bacterium, with the consequent loss of its constituents to the fluid in which it is suspended. (*see also* haemolysis)

lysogenic bacterium A bacterium that hosts a prophage, which may lie dormant even for many generations but eventually may cause lysis.

lysosomal storage disease A hereditary deficiency of one of the lysosomal enzymes involved in turnover of cellular components. An inability to degrade a substance often leads to the engorgement of the lysosomes with the undegraded material.

lysosome A vesicle that contains many hydrolytic enzymes that are active in the characteristically low-pH environment.

lytic phage (*see* bacteriophage)

M

mAb (= monoclonal antibody)

macrolide A moiety of some antibiotics (e.g. erythromycin, methymycin) which consists of a large lactone ring, synthesized largely or entirely from propionate-derived units. (*see also* polyketide; propionate rule)

macromolecule A compound or complex, usually a polymer such as a protein, nucleic acid or polysaccharide, or a covalent or non-covalent complex of any of these.

macrophage One kind of phagocytic cell.

MAD (= multi-wavelength anomalous diffraction)

magainin One of a group of anti-microbial oligopeptides secreted by the skin of frogs. (*see also* cecropin)

magic spot nucleotides Guanine nucleotide derivatives (ppGpp and pppGpp) that mediate the cessation of production of rRNA in the absence of an essential amino acid, a phenomenon present in *stringent bacteria* but absent from *relaxed bacteria*.

magnet-assisted subtractive technique (MAST) A method for detection of the tissue specificity of gene expression. Two tissues are compared: a deficient tissue from which single-stranded cDNA (driver cDNA) is prepared and attached to magnetic beads to which are also attached poly(T) sequences, and a normal tissue from which cDNA (tracer cDNA) is prepared. An excess of the driver cDNA hybridizes with the tracer cDNA preparation from all except those mRNAs in which the first tissue is deficient, and will be removed by magnetic separation of the beads. The resulting cDNA can be used to create a subtraction library for examination of tissue-dependent expression. (*see also* subtractive DNA cloning)

• Schraml, P., Shipman, R., Stulz, P. and Ludwig, C.U. (1993) Trends Genet. **9**, 70–71

magnetic circular dichroism (MCD) A technique to characterize the electronic environment of a metal ion, especially in a metalloprotein. The circular dichroism (CD) spectrum of a sample is recorded in a magnetic field parallel to the light path. Differences in the CD spectrum due to the magnetic field are interpreted as being due to Zeeman splitting of the excited state (if temperature-independent) or of the ground state, in which case much greater detail is available at very low temperatures where only one ground state is populated.

Maillard reaction (= browning reaction)

major groove The wider of the two helical spaces on the surface of an A- or B-DNA double helix. The other helical space is the *minor groove*.

major histocompatibility complex (MHC) The products of genes grouped together on a chromosome that determine whether transplanted tissues will be accepted (if from an individual with the same MHC) or rejected. In the human, they are known as *human leucocyte antigen molecules* (*HLAs*). MHC gene products are highly variable intrinsic membrane proteins of the immunoglobulin superfamily and consist of two types: *class I* form non-covalent heterodimers with β_2-microglobulin and, along with a foreign protein such as a viral antigen, comprise a recognition complex for a cytotoxic T-lymphocyte; *class II* consist of dimers of highly variable such proteins, and occur in macrophages and B-lymphocytes.

• Elliott, T., Smith, M., Driscoll, P. and McMichael, A. (1993) Curr. Biol. **3**, 854–866; Gumperz, J.E. and Parham, P. (1995) Nature (London) **378**, 245–248

MALDI-MS Matrix-assisted laser desorption/ionization mass spectrometry. (*see* mass spectrometry)

malignant Descriptive of a tumour that has a high tendency to become progressively larger and more invasive and to undergo metastasis.

manganese centre In the photosynthetic photosystem II, a complex of four manganese atoms separated by oxygen atoms that accumulates a positive charge that is used to oxidize water to molecular oxygen. (*see also* charge accumulator model)

manometry A largely obsolete technique for measurement of respiration of tissue slices or other biochemical preparations. The volume of oxygen taken up is measured in a *Warburg apparatus*, a closed system that consists of a flask with a central well in which NaOH is placed to absorb carbon dioxide. In the absence of alkali, the volume of evolved carbon dioxide is offset by oxygen taken up and is evaluated by difference. (*see also* Q_{CO_2}; Q_{O_2})

Manton–Gaulin homogenizer An apparatus for disruption of cells, especially the large quantity of bacteria or yeast that result from a fermentation, by pumping a paste of the collected cells through an orifice of adjustable size in the head of a reciprocating piston so that the shearing forces that develop break apart the cells.

MAP (= methyl-accepting protein; microtubule-associated protein)

MAP kinase Mitogen-activated protein kinase; also known as extracellular-signal-regulated protein kinase (ERK).

mapping The creation of an outline of locations of genetic markers (genes or other polynucleotide sequences) within the structures of the chromosomes. The methodology determines the resolution of the map. Being dependent upon visual markers such as break points in the chromosomes, *cytogenetic mapping* is the least well resolved. *Genetic linkage mapping* is dependent upon linkage of markers, i.e. the frequency with which markers appear together within a population. *Somatic cell hybrid mapping* is dependent upon cell lines that contain a known whole human chromosome in a rodent cell or, in the case of radiation reduction hybrids, fragments of a known human chromosome interspersed within the chromosomes of a rodent cell.

• Patterson, O. (1993) Nature Genet. **4**, 323–324; Jordan, E. and Collins, F.S. (1996) Nature (London) **360**, 111–112

marker rescue A technique to localize the site of a mutation within a gene. A polynucleotide, the 'marker', from the known wild-type sequence is allowed to recombine with the mutant genome such that, if it includes the site of the mutation, it 'rescues' the mutant by causing it to revert.

mass spectrographic immunoassay (MSIA) A technique for characterization of antigens, especially when more than one may bind to an antibody. After isolation of an antigen from a solution by using an immobilized antibody, the adsorbed antigen is eluted and analysed by mass spectrometry

• Nelson, R.H., Krone, J.R. and Williams, P. (1995) Anal. Chem. **67**, 1153–1158

mass spectrometry (MS) A technique for separation of a stream of large ions in a magnetic field. The resistance to deflection of the ion from its path is a measure of the momentum of the ion, and hence of its mass. Originally the sample material was volatilized by heat and ionized by an electron beam. Later techniques of bringing the sample out of its liquid or solid phase, i.e. *desorption*, and ionization of it, are *fast-atom bombardment MS* (*FAB-MS*) by a beam of neutral atoms, *secondary-ion bombardment MS* (*SIMS*) by a beam of ions, and *plasma-desorption MS* (*PD-MS*) by irradiation by daughter nuclides from ^{252}Cf disintegration. In *field-desorption MS* (*FD-MS*) the sample is first ionized and then volatilized, a process that induces much less thermal stress on the sample than *field-ionization MS*, in which the sample is first volatilized and then ionized in an electrostatic field. In *laser-assisted desorption MS* (*LD-MS*) and *matrix-assisted laser desorption/ionization MS* (*MALDI-MS*) the sample exchanges energy with a solid or liquid matrix in order to minimize damage to the sample and isolate one sample molecule from others. In *electrospray ionization MS* (*ES-MS*), atomization of solution particles of protein ions from which the solvent is evaporated *in vacuo* leaves multiply charged protein ions. (*see also* fast-atom bombardment mapping; fast-atom bombardment mass spectrometry; Fourier-transform mass spectrometry; ion trap; tandem mass spectrometry; time-of-flight mass spectrometry)

• Grant, E.R. and Cooks, R.G. (1990) Science **250**, 61–68; Chait, B.T. and Kent, S.B.H. (1992) Science **257**, 1885–1894; Mann, M. and Wilm, M. (1995) Trends Biochem. Sci. **20**, 219–224; Borman, S. (1995) Chem. Eng. News **73 (25)**, 23–32; Stults, J. (1995) Curr. Opin. Struct. Biol. **5**, 691–698

MAST (= magnet-assisted subtractive technique)

mast cell A cell that contains histamine-filled secretory vesicles.

mating type In effect, a sex of a haploid yeast cell that restricts it to forming a diploid cell only with a haploid of a different mating type.

mating-type switching A phenomenon in the behaviour of yeast and other organisms that rapidly change from the haploid to the diploid state. In yeast the two mating types are **a** and α; when two **a** or two α cells mate, the diploid form results in daughter cells of mating type opposite to that of the parent cells.

matrix-assisted laser desorption/ionization mass spectrometry (*see* mass spectrometry)

matrix-attachment region (*see* AT queue)

matrixin (= tissue inhibitor of matrix metalloproteinase)

matrix metalloproteinase (MMP) One of the family of Zn-containing proteins that includes fibroblast and neutrophil collagenases, gelatinase and stromelysin. (*see also* cysteine switch)

• Hooper, N.M. (1994) FEBS Lett. **354**, 1–6

maturase A nuclease involved in RNA processing; it is formed by translation of a fungal mitochondrial mRNA that is produced after a first intron is excised leaving an mRNA in which the open reading frame continues through a second intron and codes for the maturase. The enzyme, by participating in the cleaving out of the second intron, destroys its own mRNA.

Maxam–Gilbert method (chemical cleavage method) A technique for sequence analysis of DNA in which four chemical reactions are applied separately, each to cleave the polynucleotide randomly at one of the four bases. Subsequent polyacrylamide-gel electrophoresis separates the products according to chain length, and reveals the distance from the radiolabelled 3′- or 5′-end to the chemically modified base. (*see also* Sanger method)

• Maxam, A.M. and Gilbert, W. (1980) Methods Enzymol. **65**, 499–560

maximum velocity (V_{max}) The limiting rate for an enzymic reaction, shown when it is saturated with substrate. (*see also* Michaelis–Menten kinetics)

MCD (= magnetic circular dichroism)

MCP Multicatalytic proteinase. (= proteasome)

MDR (= multi-drug resistance)

mechanism A flexible term applied to the details of a process that links initial and final states, e.g. the mechanism of respiration, the mechanism of fatty acid oxidation, the mechanism of β-oxoacyl-coenzyme A dehydrogenase activity.

mechanism-based inhibitor An enzyme inhibitor that differs from the classical substrate or product analogue. One category comprises *transition-state analogues* that do not form covalent links to the enzymes but bind very tightly to them; the other category contains *enzyme-activated, k_{cat} or suicide inhibitors*, which require the action of the enzyme upon them to convert them into alkylating or other reactive agents that can form covalent complexes with the enzyme.

medial Golgi (*see* Golgi apparatus)

mediated transport (= facilitated diffusion)

medium The solid or liquid substratum on which, or in which, cells or organ explants can be made to grow; a medium may include well defined factors such as salts, amino acids and sugars, as well as less well defined factors such as serum or blood.

medium-chain fatty acid A fatty acid with a skeleton of 4–12 carbon atoms. Those with fewer carbon atoms are *short-chain fatty acids*; those with more (up to 20 carbon atoms) are *long-chain fatty acids*; and those with skeletons of 20 carbon atoms or more are *very-long-chain fatty acids*.

medulla The inner core of an organ, e.g. adrenal medulla. (*see also* cortex)

megadose An amount of a dietary supplement, such as a vitamin, mineral or essential amino acid, that is orders of magnitude greater than the officially recommended dosage.

megakaryocyte A large cell of bone marrow characterized by a multi-lobed nucleus, usually with eight times the haploid number of chromosomes; the precursor of a platelet.

meiosis The process of formation of haploid gametes from a diploid cell.

melanocyte An epithelial cell that contains melanin pigment granules, *melanosomes.*

melanosome (*see* melanocyte)

melting temperature In nucleic acid and protein chemistry, and by analogy with organic chemistry, the temperature that is the mid-point for the thermal unfolding of a nucleic acid or protein; in the case of nucleic acids, usually measured by the hyperchromic effect.

membrane asymmetry A phenomenon of membrane proteins and gangliosides involved in signal transduction and transport processes; as they are oriented towards the inside or outside, their asymmetry is appropriate to their function.

membrane potential A charge difference across a membrane due to the Donnan effect.

membrane skeleton The matrix of proteins and fibres that underlies the lipid bilayer of a plasma membrane; it contains actin filaments, actin-binding proteins and/or cell-specific structural proteins.

memory B-cell A long-lived activated B-cell that does not differentiate into a plasma cell until stimulated further.

merocrine secretion (= exocytosis)

Meselson–Radding model A conceptualization of general genetic recombination

during meiosis of newly replicated double-stranded DNA. An endonuclease nicks one strand of one daughter duplex; the new 5'-end dissociates and allows synthesis at the new 3'-end to permanently displace it. The free 5'-end displaces its homologue in the other daughter duplex, followed by excision of that displaced loop and ligation of the first 5'-end to the new 3'-end of the second duplex. Ligation of the remaining 3'- and 5'-ends and rotation to exchange the intact strands allows branch migration and separation, as in the Holliday model. (*see also* double-strand-break repair model)

• Meselson, M.S. and Radding, C.M. (1975) Proc. Natl. Acad. Sci. U.S.A. **72**, 358–361

Meselson–Stahl experiment A demonstration of semi-conservative DNA replication. *E. coli* grown on ^{15}N salts for several generations produce labelled DNA that is more dense than normal DNA, as judged by density gradient equilibrium centrifugation. When transferred to ^{14}N media, the first generation produces double-stranded DNA all of intermediate density, and the second generation DNA half of which is of intermediate density and half of normal density.

• Meselson, M. and Stahl, F.W. (1958) Proc. Natl. Acad. Sci. U.S.A. **44**, 671–682

mesenchyme The embryonic precursor of connective tissue.

***meso*-carbon** A prochiral centre; a tetrahedral carbon to which are attached two like and two other unlike groups. (*see also* Ogston hypothesis)

mesoderm One of the three primordial germ layers formed during early embryogenesis; a precursor of muscle, adipose tissue, blood vessels, the gastrointestinal tract, etc. (*see also* ectoderm; endoderm)

mesophile (*see* extremophile)

messenger RNA (= mRNA)

metabolic channelling (*see* metabolon)

metabolic pathway A sequence of metabolic reactions that transforms a substrate, e.g. glycolysis, β-oxidation, gluconeogenesis.

metabolism The sequences by which foodstuffs are degraded for the energy that is released and for transformation into cellular components.

metabolite An intermediate or end product of intermediary metabolism.

metabolon A proposed multienzyme complex that is responsible for metabolite channelling, so as to eliminate or minimize loss of efficiency due to the otherwise necessary diffusion of substrates to and of products from the enzymes of a metabolic pathway.

metal-binding finger (= zinc finger)

metalloproteinase A type of peptidase that has a metal ion at its active site. (*see also* aspartate proteinase; cysteine proteinase; serine proteinase)

metastasis The shedding of malignant cells from a tumour, which permits them to establish new colonies in other tissues.

methionine bristles The cluster of methionine side chains that are proposed to protrude from the surface of a signal recognition particle and provide a plastic hydrophobic environment that can interact with the hydrophobic residues of the signal sequence of a nascent protein.

methyl-accepting protein (MAP) One of a group of reversibly methylated proteins of some bacteria that are involved in the chemotactic response to chemoattractants.

methylation analysis Any technique to detect sites at which bases of DNA are methylated, e.g. by analysis of DNA digested by isoschizomeric pairs of restriction endonucleases.

methyl-protection footprinting (*see* footprinting)

methyl trap The accumulation of 5-methyltetrahydrofolate in vitamin B_{12} deficiency or related conditions (e.g. pernicious anaemia) due to the inability to transfer the methyl group to homocysteine by the B_{12} coenzyme-dependent transferase.

MHC (= major histocompatibility complex)

micelle A very small aggregate of matter that is dispersed in solution; often of lipoid material stabilized by detergents, or even of detergent only.

Michaelis complex (= enzyme–substrate complex)

Michaelis constant (K_m) Given by $(k_{-1} + k_2)/k_1$; the sum of the first-order rate constants for the productive and non-productive breakdown of the Michaelis complex divided by the second-order rate constant for its formation from free enzyme and free substrate; expressed in concentration units. (*see also* apparent K_m)

Michaelis–Menten kinetics The hyperbolic dependence of the rate of an enzymic reaction on substrate concentration that can be analysed mathematically according to a model that requires the reaction to proceed through a reversibly formed enzyme–substrate complex, followed by conversion of the complex into product; contrasted with sigmoidal dependence of rate on substrate concentration.

microbody (= peroxisome)

microbore HPLC (*see* high-pressure liquid chromatography)

microcluster An aggregate of several (2–10) membrane receptors that forms in response to cell stimulation; such an aggregate is a transitory state between the receptors being dispersed in the membrane and their formation into much larger aggregates, patches that will undergo endocytosis.

• Hollenberg, M.D. (1991) FASEB J. **5**, 178–186

microheterogeneous (*see* monodisperse)

microreversibility (= microscopic reversibility)

microsatellite One of many short, highly polymorphic, non-coding sequences, e.g. poly(TG), found well spaced throughout the genome that can serve as landmarks during physical mapping.

• Beckman, J.S. (1990) Science **248**, 18

microscopic reversibility A principle from molecular physics that holds that each proposed step of a catalytic cycle must in theory be reversible, and particularly that the forward and reverse reactions must pass through the same transition state.

microsome An artifactual vesicle formed from the endoplasmic reticulum when cells are disrupted.

microtitre plate A transparent tray with 96 or 384 wells in which chromogenic reactions may be run and then automatically scanned to record changes in absorbance.

microtubule A large (30 nm diameter) rigid component of the cytoskeleton that is built of α- and β-tubulin subunits and associated proteins and serves as a component of cilia, eukaryotic flagella and mitotic spindles.

microtubule-associated protein (MAP) One of the non-tubulin proteins that are associated with microtubules in some of their cellular sites.

microtubule-organizing centre (MTOC) A cellular structure that serves as a site for nucleation of microtubule growth. (*see also* centriole)

microvillus A finger-like projection from the apical membrane of some epithelial cells, especially those of the intestine.

mimetic A non-peptide synthetic molecule that reproduces some features of a natural peptide, e.g. it may bind strongly to an antibody by placement of the same functional groups, in the same relation to each other, as are found in the peptide antigen.

• Saragovi, H.U., Fitzpatrick, D., Rakatabutr, A., et al. (1991) Science **253**, 792–795

mimotope A synthetic peptide that binds to an antibody against an unidentified antigen; usually a peptide selected from a large library of peptides. (*see also* random peptide library)

• Scott, J.K. and Smith, G.P. (1990) Science **249**, 386–390

mineralocorticoid An adrenal steroid that promotes resorption of Na^+ by the kidney tubule; notably, aldosterone.

miniantibody (= single-domain antibody)

minichromosome A circular double-stranded DNA structure packaged with protein into nucleosomes or higher-level organization of chromatin. (*see also* double minute chromosome)

minimal molecular mass A value for the molecular mass of a macromolecule calculated from its content of a metal ion, end group or other feature present in an unknown stoichiometric ratio, e.g. the minimal molecular mass of haemoglobin calculated from its iron content is: 55.84 g/g-atom of iron $\times$ 100/0.34% iron in haemoglobin = 16 400 g/mol.

minimum daily requirement The quantity of nutrients required each day in the diet by average individuals. (*see also* recommended daily allowance)

minisatellite variant repeat-specific priming PCR (MVR-PCR) A method for characterization of DNA samples of individuals for forensic purposes, based on sequence variations within the repeating units of variable number tandem repeats (VNTRs). PCR primers are designed for a common flanking region and each of the variant repeating units; the latter primers are initially used in low enough concentration so that on average only one binds per template; thus upon amplification and electrophoresis, a ladder is produced for each primer pair that indicates the position of the variant along the length of the VNTR. The sequence of the variants derived from ladders for all primer pairs is subject to great diversity, and is both characteristic of an individual and capable of digitalization for data storage, search and retrieval.

- Jeffreys, A.J., MacLoed, A.J., Tamaki, K., Neil, D.L. and Monckton, D.G. (1991) Nature (London) **354**, 204–209

minor groove (*see* major groove)

minus-strand (*see* plus-strand)

mismatch analysis (= heteroduplex analysis)

mismatched PCR A method for detection of a point mutation that neither produces nor destroys a restriction site; intended especially for screening large numbers of individuals. One of the PCR primers is designed so that it is complementary to the sequence immediately upstream (or downstream) from the mutation, and near its 5'-end (or 3'-end) contains a mismatched base. The mismatch and the nearby mutated base (but not the normal base) when amplified generate a new restriction site. The amplified mutant DNA (unlike the amplified normal DNA), when digested with appropriate endonucleases, electrophoresed and probed with an appropriate labelled oligonucleotide, will reveal a new, smaller restriction fragment. Alternatively, the mismatched primer may be designed to create a new restriction site with the normal, but not the mutant, DNA. (*see also* restriction-site-generating PCR)

missense mutation A mutation that substitutes one amino acid residue for another.

missense suppressor A mutation that produces a tRNA that can recognize a missense mutation and insert into the protein sequence an amino acid residue that permits phenotypic expression of the gene.

mitochondrion An organelle of eukaryotic cells that is the site of oxidative phosphorylation, the tricarboxylic acid cycle, β-oxidation, etc., and is composed of a relatively permeable outer membrane and a deeply indented inner membrane.

mitogen A compound that induces cell division.

mitoplast A preparation made from mitochondria in which the outer membrane is stripped away by detergents and the inner membrane with the enclosed matrix is left intact.

mitosis Cell division that produces daughter cells with the same number of chromosomes as the parent cell.

mitosol The soluble matrix that occupies the space inside the mitochondrial inner membrane.

mixed disulphide An oxidation product of two dissimilar thiol compounds.

mixed-function oxidase (mono-oxygenase) An enzyme that reduces molecular oxygen by incorporation of one atom into water and the other into another substrate; e.g. phenylalanine hydroxylase. (*see also* dioxygenase)

mixed inhibition (= non-competitive inhibition)

mixotrophic In microbiology, descriptive of an organism that requires organic material and light for growth. (*see also* autotrophic; heterotrophic)

MMP (= matrix metalloproteinase)

Mn^{2+} quench method A method for observation of the control of bivalent cation flux across the plasma membrane of a cell. Mn^{2+} is used in place of Ca^{2+} because, unlike Ca^{2+}, it does not accumulate in intracellular sites. The entry of Mn^{2+} into a cell is monitored by the quenching of fluorescence of a fluoro-phore (fura-2) with which the cell has been loaded.

- Jacob, R. (1990) J. Physiol. (London) **421**, 55–77

mobile barrier A proposed mechanism of membrane transport. A transmembrane protein, the activated porter, exposes its binding site for the transported molecule to one membrane face and then to the other. In the case of a *uniport porter*, activation does not depend upon the solute, so transport can be effected in either direction; a *symport porter* is activated only when both (or neither) of the two solutes are bound; an *antiport porter* is activated when either, but not both, of the two solutes is bound. This proposal is often coupled with the concept of *solvation substitution*, the identification of solute binding as the replacement of bound water by specific liganding groups of the porter, e.g. chelation of ions, to help explain the specificity of the porter and how it maintains a barrier against transport of water across the membrane. (*see also* mobile carrier)

mobile carrier A proposed mechanism of membrane transport similar to the mobile barrier mechanism, but in which the porter is imagined to be a revolving door that accepts one (uniport) or two (symport) solutes, or distinguishes between membrane faces (antiport).

mobile genetic element A transposon or an insertion sequence; a polynucleotide sequence that can move from a chromosome or plasmid to another chromosome or plasmid. (*see also* intron homing; protein splicing)

mobility shift assay (= gel shift assay)

model substrate A synthetic substrate that often contains only the minimum necessary features to satisfy the specificity of the enzyme, but sometimes incorporates chemical features to facilitate detection of enzyme activity; e.g. a *p*-nitrophenyl ester as an endopeptidase substrate, as the *p*-nitrophenolate product is intensely coloured.

module In protein chemistry, a building block from which larger proteins are constructed; a recurring folding pattern in proteins, examples of which may have amino acid sequences that are homologous or non-homologous, e.g. the kringle, the EF hand.

• Baron, M., Norman, D.G. and Campbell, I.D. (1991) *Trends Biochem. Sci.* **16**, 13–17

moiety A fragment of a molecule, especially one that comprises an identifiable unit, e.g. an acetyl or pyridoxal phosphate group, a regulatory subunit.

molar absorption coefficient (extinction coefficient) The (calculated) absorbance of a 1 M solution of a chromophore present in a 1 cm light path. (*see also* Beer–Lambert equation)

molecular accordion A reporter molecule; a fluorophore-labelled linear polymer that interacts with natural and synthetic phospholipid membranes, and whose fluorophores respond to the polymer's physical environment. The polymer attaches to the membrane through its aromatic fluorophores, which are sparsely spaced along its length and allow the intervening polymer to loop out into the solvent. The fluorescence intensity is greatly enhanced by the mutual proximity of the fluorophores, so physical forces that promote close association of the polymer loops (e.g. a lowered temperature or a solvent that interacts poorly with the polymer) will close up the loops, draw the fluorophores closer together and increase fluorescence, while a strongly interacting solvent or a higher temperature will cause the loops to relax, separate the fluorophores, and decrease fluorescence.

• Cates, M. (1991) *Nature* (London) **351**, 102

molecular archaeology (*see* jumping PCR)

molecular channel A proposed feature of multienzyme complexes; the product of catalysis by the first enzyme of the metabolic sequence travels through the channel to the active site of the next enzyme, etc.

molecular chaperone (*see* chaperone machine)

molecular chronometer A phylogenetic marker; a highly conserved protein (e.g. ubiquitin) or nucleic acid (e.g an rRNA) whose rate of mutation is constant, and which can therefore be used to construct phylogenetic trees.

• Potter, S. (1992) *The Biochemist* (Bull. Biochem. Soc.) **14(2)**, 16–20

molecular clock hypothesis The proposal, so far experimentally unverified, that mutations have occurred at random and at a constant frequency in all species throughout the billions of years of biochemical evolution. Successful mutations, i.e. those that persist, are only those that do not adversely affect stability or function.

• Behe, M.J. (1990) Trends Biochem. Sci. **15**, 374–376

molecular dynamics The computer-simulated quantitative evaluation of a molecular model by assignment of force constants to bonded and non-bonded interactions between neighbouring atoms so that bond distances and angles are confined to an acceptable range and crowding is minimized; often coupled with *energy minimization*, which makes small adjustments in the locations of atoms to optimize these interactions.

molecular evolution The changes in gene polynucleotide sequences and resulting protein structures that have paralleled the evolution of species.

molecular filter The function of specificity in membrane protein trafficking that directs membrane components to their appropriate locus; a phenomenological, rather than a morphological, designation.

• Koch, G. and Hopkins, C.R. (1992) Trends Biochem. Sci. **17**, 211–212

molecular imprinting The formation of very specific binding, or even catalytic, characteristics of a matrix by polymerization of a suitable monomer in the presence of a ligand or a transition-state analogue. (*see also* bio-imprinting)

• Vidyasankar, S. and Arnold, F.H. (1995) Curr. Opin. Biotechnol. **6**, 218–224

molecular matchmaker A protein that effects complex-formation between another protein and DNA by catalysis of an energy-dependent conformational change in at least one of the pair; it thereby increases affinity and permits binding that would not occur spontaneously. (*see also* chaperone machine)

• Sancar, A. and Hearst, J.E. (1993) Science **259**, 1415–1420

molecular mechanics (= molecular dynamics)

molecular model A physical representation, e.g. balls and sticks, used to represent the structure of a compound or macromolecule, or a computer graphics simulation of such a structure. More sophisticated versions test structure–function relationships by simulation of ligand binding and conformational changes (*see also* docking; molecular dynamics)

molecular modelling Any of the computational approaches for relating chemical or physical properties to structure, e.g. design of agonists or antagonists from the X-ray crystallography data of a receptor protein.

molecular pharming The production of human proteins of pharmacological interest in the milk of transgenic farm animals.

• Bialy, H. (1991) BioTechnology **9**, 786–788

molecular ruler A means of measurement of distances on a molecular scale; the estimation of distances between a fluorescence energy donor and receptor, evaluated from the efficiency of resonance energy transfer, which decreases according to the sixth power of the distance that separates them. During energy transfer, excitation of the donor results in *energy transfer* to the acceptor, which fluoresces. (*see also* resonance energy transfer)

mole percent excess A quantitative measure of the concentration of a stable isotope, analogous to the specific radio-activity of a radioisotope; the enrichment of that isotope, as a percentage of all isotopes, over and above its usual occurrence in nature.

Molisch reaction A colorimetric method for estimation of sugars that involves treatment with sulphuric acid and α-naphthol.

molten globule intermediate A stable intermediate between the denatured and native states of a protein that is compact and displays considerable secondary structure, but allows amino acid side chains, especially hydrophobic side chains, to be exposed to the bulk solvent phase. (*see also* framework model; two-state rule)

• Baldwin, R.L. (1990) Nature (London) **346**, 409–410; Baldwin, R.L. (1991) Chemtracts Biochem. Mol. Biol. **2**, 379–389; Bychkova, V.E. and Ptitsyn, O.B. (1993) Chemtracts Biochem. Mol. Biol. **4**, 133–163

mono- A prefix that indicates one basic unit, without any polymerization, as in mononucleotide, monosaccharide. (*see also* oligo-; poly-)

monocistronic mRNA An mRNA that codes for a single protein; characteristic of eukaryotes. (*see also* cistron; operon; polycistronic mRNA)

monoclonal antibody (mAb) An immunoglobulin preparation that is completely homogeneous, due to its formation by daughters of a single progenitor cell that has been programmed for the synthesis and secretion of one specific antibody. (*see also* polyclonal antibody)

monocyte One type of non-granular leucocyte; a precursor of macrophages and osteoclasts.

monodisperse Descriptive of a population of polymers in which all individual molecules have the same covalent structure and molecular size. It is contrasted with *polydisperse*, descriptive of a population of polymers that have the same repeating unit(s) and the same covalent structure, but vary in molecular size. *Paucidisperse* describes a population of related polymers that share the same basic covalent structure, and perhaps function, but show a low level of dispersity. *Microheterogeneous* describes a population of polymers with the same primary structure, and possibly the same molecular size, but with limited modifications of some monomeric units.

Monod–Wyman–Changeux (MWC) model A concerted model for the basis of co-operativity in multimeric binding proteins, originally presented as a description of oxygen binding to each of the haem groups of the haemoglobin tetramer. In the absence of a ligand (oxygen in the case of haemoglobin) the subunits exist in a low-affinity '*taut*' *conformation* (tense form). The ligand may bind to as many as two monomers in this conformation, at which point all four monomers convert from the low-affinity form to the high-affinity '*relaxed*' *conformation* (R-form). (*see also* Koshland model)

mono-oxygenase (= mixed-function oxidase)

monoterpene (*see* terpene)

morgan (*see* centimorgan)

morphogen A molecule, secreted by a tissue, that acts by a concentration-dependent mechanism to induce position-specific patterning of distant (non-adjacent) cells during early embryogenesis.
- Johnson, R.L. and Tabin, C. (1995) Cell **81**, 313–326

mosaicism The expression of a mutation in some somatic cells but not others of an organ or tissue.

Mössbauer spectroscopy A technique for studying the chemical environment of some metal atoms in solid samples, especially ^{57}Fe. The metal is irradiated with soft γ-radiation and emits lower-energy γ-radiation that is indicative of its valence state, electronic and magnetic environment.
- Stevens, J.G., Bowen, L.H. and Whatley, K.M. (1988) Anal. Chem. **60**, 90R–106R

motif A recurring pattern of protein supersecondary structure, e.g. immunoglobulin fold, kringle, zinc finger.

motor In cell biology, a complex of structures responsible for intracellular movement, e.g. the microtubule-anchored spindle motor, the actomyosin engine.
- Gelfand, V.I. and Scholey, J.M. (1992) Nature (London) **359**, 480–481; Cooke, R. (1995) FASEB J. **9**, 636–642; Hasson, T. and Mooseker, M.S. (1995) Curr. Opin. Cell Biol. **7**, 587–594

M period (*see* cell cycle)

mRNA Messenger RNA; the RNA that contains the coded information, as sequences of codons, for protein synthesis.

msDNA (= multiple-copy single-strand DNA)

MSIA (= mass spectrographic immunoassay)

MS/MS (= tandem mass spectrometry)

MS^n (*see* tandem mass spectrometry)

mtDNA Mitochondrial DNA; a double-stranded polynucleotide composed of a heavy and a light chain, distinguished by the buoyant densities of the separated strands, which are determined by their G + T content.

MTOC (= microtubule-organizing centre)

mucin An extracellular glycoprotein with a high carbohydrate content that contains many serine and threonine residues bearing the O-linked carbohydrate moieties.

mucopolysaccharide (= proteoglycan)

mucosa The external secretory and absorptive surface of the gastrointestinal tract.

multicatalytic proteinase (= proteasome)

multi-dimensional MS (*see* tandem mass spectrometry)

multi-drug resistance (MDR) A phenomenon in which resistance that develops to one drug is manifested as resistance to many others, even those with very different structures; seen especially in chemotherapy. MDR is due to traffic ATPase transporters of very low specificity that extrude drugs or sequester them in vesicles.

multienzyme complex An aggregate of enzymes of specific composition and geometry that act in sequence on a substrate and that, although separable *in vitro*, fit and function together for efficiency, e.g. pyruvate dehydrogenase complex, bacterial fatty acid synthase.

multigene family A group of genes that code for homologous products with similar functions, e.g. the genes for globins. (*see also* superfamily)

multimeric Descriptive of a protein composed of several subunits.

multiple-copy single-strand DNA (msDNA) A satellite DNA detected in some bacteria that features an oligoribonucleotide attached by the 2′-hydroxy group of an internal guanylate residue to the 5′-end of an oligodeoxyribonucleotide, while the 3′-end of the DNA moiety is base-paired with the 5′-end of the RNA moiety; both the DNA and RNA possess considerable internal base-pairing.
• Inouye, M. and Inouye, S. (1991) Trends Biochem. Sci. **16**, 18–21

multiplex amplification (= multiplex PCR)

multiplex fluoresence *in situ* hybridization (multiplex FISH) A method involving the use of several fluorophore-tagged oligonucleotides to simultaneously label metaphase chromosomes, and their discrimination and visualization through the use of narrow band filters and computerized output.
• Le Beau, M.M. (1996) Nature Genet. **12**, 341–344

multiplex PCR A screening technique for any of several mutations known to cluster in a relatively small number of sites (hot spots) in a gene, especially microsatellites. Simultaneous PCR reactions with several restriction fragments and the same number of pairs of primers designed to flank the deletion-prone hot spots give as many amplified fragments when normal cells or tissues are scanned, but a mutant will lack one of the restriction fragments. As the fragments sort into clusters of closely similar sizes, there is no overlap of fragments due to restriction at different sites.
• Chamberlain, J.S., Gibbs, R.A., Ranier, J.E. and Caskey, C.T. (1990) in PCR Protocols: A Guide to Methods and Applications (Innis, M.A., Gelfand, D.H., Sninsky, J.J. and White, T.J., eds.), pp. 272–281, Academic Press, San Diego

multivesicular body (MVB) A subcellular structure that travels between the inside surface of the plasma membrane and endosomes, perhaps along a tubular endocytotic network. It scavenges endocytosed receptors destined for degradation, delivers them to endosomes and facilitates the return to the plasma membrane of receptors destined for recycling.
• Warren, G. (1990) Nature (London) **346**, 318–319

multi-wavelength anomalous diffraction (MAD) For proteins that contain a heavy metal, or synthesized with selenomethionine in place of methionine, an approach to solving their X-ray crystallography structure that takes advantage of the anomalous diffraction, in addition to the classical diffraction, that occurs when the crystal is irradiated by synchrotron X-rays of defined energies on the absorption edge of the heavy atoms. The additional diffractions provide sufficient information for assignment of the phases and amplitudes of the crystal's diffraction. (*see also* isomorphous replacement; phase problem)
• Geisow, M. (1991) Trends Biotechnol. **9**, 4–5

multiwire data collector A variety of area detector; an array of individual proportional photon detectors individually wired to a computer to allow quantification of energy and, in crystallographic experiments, its angle of diffraction.

mutagen A compound that causes a mutation by chemical modification of a base of DNA or as an intercalating agent

mutarotation The change in optical rotation that accompanies the approach to equilibrium of the two anomers of a sugar from a solution of one anomer alone; also the conversion of one anomer of a sugar into the other.

mutein A mutant protein, especially a product of recombinant DNA technology.

MVB (= multivesicular body)

MVR-PCR (= minisatellite variant repeat-specific priming PCR)

MWC model (= Monod–Wyman–Changeux model)

mycoplasma A type of bacteria that has no cell wall.

myeloma cell A bone-marrow tumour cell that produces and secretes an immunoglobulin.

myo- A prefix that indicates muscle, e.g. myocyte.

myoblast An immature muscle cell.

myokinase A former name for adenylate kinase.

myoneural junction (= neuromuscular junction).

N

naive In immunology, descriptive of an unimmunized animal or a memory T-cell that has not been activated to respond to an antigen.

naptonuon (*see* junk DNA)

narrow-bore HPLC (*see* high-pressure liquid chromatography)

NASBA (= nucleic acid sequence-based amplification)

nascent Descriptive of a newly formed molecule during its formation or at the moment of its completion.

native gel A polyacrylamide gel for support of electrophoresis of proteins that is formulated without sodium dodecyl sulphate, so that the protein remains in its native form. (*see also* sodium dodecyl sulphate/polyacrylamide-gel electrophoresis)

native structure The structure of a protein or nucleic acid in which it is able to perform its physiological function. (*see also* denaturation)

natural killer (NK) cell A cytotoxic lymphocyte that recognizes foreign cells but, unlike the killer T-cell, need not be previously sensitized to act.

• Gumperz, J.E. and Parham, P. (1995) Nature (London) **378**, 245–248

natural product (*see* secondary metabolism)

natural selection A hypothetical explanation of evolution, that successful organisms, over time, will crowd out and displace less successful ones, or will be better adapted to different niches and, over

time and by genetic drift, lead to new species.

nearest-neighbour analysis A method of analysis in which a polymer is first synthesized from a precursor labelled in a specific atom, then degraded in such a way that the labelled atom appears with the adjacent moiety and allows deduction of the identity of the pair that had shared that atom in the polymer.

necrosis (*see* apoptosis)

negative co-operativity A form of allosteric behaviour of multimeric enzymes, in which binding of substrate to one subunit decreases the affinity of the substrate for other subunits. (*see also* positive co-operativity)

negative energy balance The dietary situation in which caloric dietary intake is smaller than energy expenditure, resulting in loss of weight. (*see also* positive energy balance)

N-end rule The observation that the half-life of a protein *in vivo* is a function of its N-terminal residue. (*see also trans*-recognition)

• Varshavsky, A. (1992) Cell **69**, 725–735

neoglyco- A prefix denoting a synthetic complex carbohydrate moiety that has been attached to a protein or lipid.

neoplasm (= tumour)

nephelometric Turbidimetric.

Nernst equation A quantitative expression of the relationship of concentration differences of an ion across a membrane: $\Delta\Psi = (\mathbf{R}T/n\mathcal{F})(\ln [X^n]_e/[X^n]_i)$, where $\Delta\Psi$ is the membrane potential, $\mathbf{R}$ the gas constant, T the temperature, n the ionic charge, $\mathcal{F}$ the Faraday constant, $[X]_e$ the external concentration of the ion and $[X]_i$

the internal concentration. Also an expression of the redox potential of a system: $E = E_o + (RT/n\mathcal{F})\ln([ox]^n/[red]^n)$, where E and E_o are respectively the actual and standard electrical potentials of the system, and [ox] and [red] are the concentrations of the oxidized and reduced species respectively.

nerve gas Often a fluorophosphate inhibitor of acetylcholinesterase; chemically and functionally related to organophosphate insecticides.

nested deletion sequencing A method for extension of the length of a target DNA that can be sequenced using a single primer. Removal by an appropriate nuclease of varying polydeoxynucleotide lengths from the region near the primer binding site on a plasmid brings into sequencing range different regions of the target, followed by religation and circularization of the resultant shortened plasmids.

• Rashtchian, A. (1995) Curr. Opin. Biotechnol. **6**, 30–36

nested primers A second set of PCR primers whose use follows an initial copying of a nucleic acid sequence; the nested primers are specific for sequences internal to the 3'- and 5'-ends of the originally copied sequence.

Neuberg ester Obsolete name for fructose 6-phosphate.

neurofilament An intermediate filament seen microscopically in neurons that may consist of one of several types of protein.

neuromuscular junction (myoneural junction) The attachment site of a neuron on a muscle cell, analogous to a synapse.

neuron A nerve cell.

neurophysin A protein associated with oxytocin or vasopressin in the secretory cells of the neurohypophysis and in blood; derived from the same protein precursor, proneurophysin, as the hormone it binds.

Neurospora crassa A bread mould; a subject of much experimentation. (*see also* one gene one enzyme hypothesis)

neurotransmitter A chemical signal that passes from an afferent nerve across the synapse to stimulate the efferent nerve, e.g. acetylcholine, noradrenaline (norepinephrine), dopamine.

neutral glyceride An uncharged fatty acid ester of glycerol; i.e. a mono-, di- or triacylglycerol, as contrasted with a phospholipid.

neutron diffraction A crystallography technique that uses neutrons (rather than the X-rays of more conventional crystallography) which are scattered by nuclei and can therefore locate hydrogen atoms.

• Kossiakoff, A. (1986) Methods Enzymol. **131**, 433–447

neutrophil (= polymorphonuclear leucocyte)

nicking As contrasted with cutting, the cleaving of a single, or very few, phosphodiester bonds in one polynucleotide strand of a DNA duplex; also the cleavage of a peptide bond in a protein that does not denature the protein or release a peptide. (*see also* cut)

nick-translation A technique often used for introduction of ^{32}P into a DNA duplex involving cleavage of a phosphodiester bond of one strand, excision of the newly exposed nucleoside 5'-phosphate residue, and introduction of a new nucleotide from a radiolabelled triphosphate substrate.

NIH shift The mechanism of a step in the catabolism of tyrosine, named for the National Institutes of Health, where it was described. The sequence of events by which *p*-hydroxyphenylpyruvate hydroxylase forms homogentisic acid involves α-oxidative decarboxylation to form a peroxyacid, epoxidation of the aromatic ring by the peroxyacid, and migration of the side chain to the site on the ring adjacent to its original position.

ninhydrin reaction A colorimetric method for quantification of amino acids; reaction with ninhydrin oxidizes and decarboxylates amino acids and generates an intense blue chromophore from reduction and coupling of the reagent with itself.

Nissl substance (= rough endoplasmic reticulum)

nitrogen balance The quantitative difference between the intake of all nitrogenous products in the diet and the excretion of all nitrogenous products (both expressed in terms of g of nitrogen); *positive nitrogen balance* if there is a net uptake of nitrogen and *negative nitrogen balance* if a net loss.

nitrogen cavitation A technique for disruption of cells prior to their fractionation. A suspension of cells is subjected to high nitrogen pressure, which is suddenly released so as to cause the cells to burst as nitrogen bubbles form inside them.

nitrogen fixation The reduction of molecular nitrogen to ammonia and nitrogen-containing metabolites that is carried out synthetically, or naturally by blue–green algae or bacteria, e.g. *Rhizobium* species that inhabit nodules on the roots of legumes.

nitrogen mustard An alkylating agent that can cross-link adjacent guanine bases of DNA and thus interfere with its function, e.g. $CH_3N(CH_2CH_2Cl)_2$ (mechlorethamine).

nitrous acid method (*see* Van Slyke method)

NK cell (= natural killer cell)

N-linked carbohydrate A carbohydrate moiety of a glycoprotein that is attached by a glycosidic bond to the β-amide nitrogen of an asparagine residue; may be complex-type or high-mannose-type. (*see also* O-linked carbohydrate)

NMR (= nuclear magnetic resonance)

node The point of contact of a strand of supercoiled DNA where it loops back on itself. A node, and the *writhe* of the DNA, may be positive or negative; it is positive if, when viewed so that the strand enters the node from below and leaves from above, and the shortest turn from strand entry and strand exit is clockwise. Negative nodes are formed from negatively supercoiled DNA. A *knot* in the DNA is seen when three nodes all of the same sense, e.g. all negative, occur together; this is seen in electron micrographs as a *trefoil* structure, a knot from which extend three DNA loops. (*see also* intasome)

- Nash, H.A. (1990) Trends Biochem. Sci. **15**, 222–227

NOE Nuclear Overhauser effect. (*see* nuclear Overhauser and exchange spectroscopy)

NOESY (= nuclear Overhauser and exchange spectroscopy)

nomogram The presentation of three parallel linear or non-linear scales of related values aligned so that a straight line between desired points on two of the scales also crosses the third scale at the corresponding value, e.g. a line through points on scales for centrifuge rotor diameter (cm) and speed (rev./min) also passes through the scale for centrifugal force (*g*) at the correct point.

non-coding strand (*see* antisense)

non-collagen collagen A protein complex that, like collagen, has (Xaa-Yaa-Gly)$_n$ amino acid sequences and non-helical globular regions. The collagen-like sequences assemble together into a triple helix and the globular regions project out in a '*flower bouquet*' pattern; e.g. acetylcholinesterase, complement protein C1g.

non-competitive inhibition (mixed inhibition) A form of enzyme inhibition in which the inhibitor binds to both the free enzyme and the enzyme–substrate complex, resulting in an increase in K_m and a decrease in V_{max}. (*see also* competitive inhibition; inhibitor; uncompetitive inhibition)

non-essential amino acid (*see* essential amino acid)

non-haem iron protein (= iron–sulphur protein)

non-overlapping Descriptive of the genetic code, in which the three bases of one codon are distinct from the bases of adjacent codons.

non-permissive host A cell that does not allow the multiplication within it of a virus. (*see also* permissive host)

non-productive In enzymology, descriptive of an enzyme–substrate complex other than the Michaelis complex; one that cannot be converted into an enzyme–product complex. In protein chemistry, descriptive of a partially folded conformer that is not on a folding pathway between a random coil and the native state. (*see also* framework model)

non-reducing sugar A sugar whose anomeric carbon atom forms part of a glycosidic bond, rendering it unable to be oxidized by an alkaline cupric ion solution. (*see also* reducing sugar)

nonsense mutation A mutation that generates one of the three chain termination codons. (*see also* amber mutation; ocher mutation; opal mutation)

non-synonymous (*see* synonymous)

non-transcribed strand (*see* antisense)

nor- A prefix that signifies a product of the removal of a -CH$_2$- or CH$_3$ group, e.g. noradrenaline, a 19-norsteroid.

northern blotting By analogy with Southern blotting, a technique for detection of specific RNA molecules; RNA from a cell is denatured and separated by slab gel electrophoresis, then blotted on to a sheet of nitrocellulose or nylon and hybridized with radiolabelled DNA that is complementary to the desired RNA, whose presence is subsequently indicated by autoradiography. (*see also* electroblotting; Southern blotting; southwestern blotting; western blotting)

N protein (*see* anti-terminator)

N-terminal In a polypeptide sequence, that unique residue which is connected to the linear sequence by its carboxy group, leaving it with a free amino group. In practice, the amino group of an N-terminal residue may be modified, e.g. by acylation by an acetyl or fatty acyl group. (*see also* C-terminal)

NTP Nucleoside triphosphate; the letter N refers to any or all of the common bases.

nuclear envelope The structure that encloses the eukaryotic nucleus and consists of an inner and an outer nuclear membrane and the perinuclear space between the two.

nuclear localization signal A polypeptide sequence of a protein that permits its entry through nuclear pores into the nucleus of the cell.

nuclear magnetic resonance (NMR) A technique for studying the microenvironment of certain atoms that can absorb energy in a magnetic field, e.g. ^{1}H, ^{13}C, ^{15}N, ^{31}P. NMR can be used to study the conformation in solution of polypeptides, proteins and other molecules.

• Jardetzky, O. and Schmitt, T.H. (1996) Chemtracts Biochem. Mol. Biol. **6**, 1–17

nuclear Overhauser and exchange spectroscopy (NOESY) A nuclear magnetic resonance method for evaluation of internuclear distances [within 4 Å (0.4 nm)] between non-covalently bonded nuclei; one centre is saturated with radiation and, as its excited state decays, the effect on the spin distribution of the neighbouring centre (the *nuclear Overhauser effect*) is observed. (*see also* correlation spectroscopy)

• Kay, L.E. (1995) Curr. Opin. Struct. Biol. **5**, 674–681

nuclear pore A gap in the nuclear envelope that provides a selective barrier to entry or exit of macromolecules.

nuclear receptor A nuclear protein that allows regulation of gene expression by a lipophilic effector molecule, such as a steroid or thyroid hormone, eicosanoid, retinoid or ecdysone. When the effector molecule is bound to the receptor protein, the latter can also bind to a specific DNA sequence. (*see also* cytosolic receptor)

• Manglesdorf, D.J., Thummel, C., Beato, M., et al. (1995) Cell **83**, 835–839

nuclear run-off assay A method to determine which genes in a population of cells are expressed at a given time. Nascent RNA transcripts are radiolabelled with [^{32}P]NTPs as they are elongated in isolated nuclei (new transcripts are not initiated in these isolated nuclei). To determine if a specific gene is expressed, the radiolabelled run-off transcripts are used as hybridization probes with DNA from recombinant plasmids that contain the gene of interest. (*see also* run-on assay)

nuclear scaffold The network of protein fibres that underlies the nuclear inner membrane and is exposed upon extraction of soluble proteins from the nucleus. Between mitoses (interphase), chromatin is attached to the scaffold at points along its DNA strands. The loops of unattached DNA are presumably available for transcription.

nucleation In the process of folding of a protein, the rate-limiting formation of the first elements of secondary or tertiary structure around which the remainder of the protein subsequently folds.

nucleic acid sequence-based amplification (NASBA) A method for amplification (up to 10^9-fold) of a single-stranded RNA template. A reaction mixture contains RNA polymerase, ribonuclease H, reverse transcriptase, nucleoside triphosphates and two polydeoxynucleotide primers; primer 1 contains a promoter for the RNA polymerase and a sequence complementary to the 5'-end of the RNA template's coding region, and primer 2 is the DNA analogue of the 3'-end of the template. The amplification begins as primer 1 anneals to the template, the reverse transcriptase forms a cDNA:RNA hybrid, the RNA of which is hydrolysed by ribonuclease H. With formation of the resulting single-stranded (ss)DNA, the reaction enters a cyclic phase: the ssDNA anneals to primer 2, and the reverse transcriptase then synthesizes a complementary copy of the first-synthesized DNA to generate a double-stranded promoter region which the polymerase recognizes and uses to make 10–100 RNA transcripts. As each new RNA transcript is a template for the reverse transcriptase, the cycle continues to amplify the original RNA template.

• Compton, J. (1991) Nature (London) **350**, 91–92

nucleofuge A chemical group that is displaced in a nucleophilic substitution or elimination reaction. (*see also* electrofuge)

nucleolus A structure in the nucleus of a cell that is the site of synthesis of rRNA.

nucleophile A compound or functional group that can attract and bind a proton or another positively charged species such as a carbocation ion. (*see also* electrophile)

nucleoplasm Between mitoses (interphase) the visibly unstructured matrix of the nucleus, which consists of all but the nucleolus.

nucleoprotein A protein–DNA or protein–RNA complex.

nucleoside A purine or pyrimidine base with a sugar attached in a glycosidic linkage; *ribonucleoside* if the sugar is D-ribose and *deoxyribonucleoside* if the sugar is D-deoxyribose. (*see also* nucleotide)

nucleosome The repeating structural unit of chromatin that consists of a complex of eight molecules of histones and a DNA double helix wrapped twice around them.

nucleotide A nucleoside with one, two or three phosphate(s) esterified to one of the free hydroxy groups of the sugar moiety, usually the 3'- or 5'-hydroxy group; *ribonucleotide* if the sugar is D-ribose; *deoxyribonucleotide* if the sugar is D-deoxyribose. (*see also* nucleoside)

nucleotide-binding domain One of the homologous supersecondary structures of some proteins, composed of Rossmann folds, to which may be bound the AMP moiety of a substrate or ligand; e.g. the domain of a dehydrogenase that binds a pyridine nucleotide or a flavin–adenine dinucleotide.

nucleotide-binding fold (= nucleotide-binding domain)

nucleus A membrane-limited structure of eukaryotic cells that contains chromatin and is the site of DNA and RNA synthesis.

nuon (*see* junk DNA)

nut site N utilization site. (*see* anti-terminator)

O

O-antigen (*see* lipopolysaccharide)

occupancy The degree to which a binding site, e.g. a receptor, is liganded, e.g. by its hormone.

ocher mutation A nonsense mutation that, due to the premature appearance in the mRNA of the terminator codon UAA, leads to the formation of a truncated, possibly non-functional, protein. (*see also* amber mutation; opal mutation)

ODN Oligodeoxynucleotide

oestrogen (estrogen) A compound, usually a steroid, that supports the development of female secondary sex characteristics; e.g. oestradiol.

O'Farrell gel (= two-dimensional electrophoresis)

Ogston hypothesis The proposal that an enzyme can distinguish in its substrate between two like substituents on a carbon atom that also carries two unlike groups, through binding sites for three of the four substituents, which forces the substrate to display in a unique configuration one of the like groups. (*see also* chirality; *meso*-carbon; symmetry)

• Ogston, A.G. (1978) Nature (London) **162**, 963

Okazaki fragment Fragments of a single-stranded DNA synthesized on the discontinuous site of a DNA replication fork. (*see also* semi-discontinuous)

OLA (= oligonucleotide ligation assay)

oligo- A prefix that indicates a moderate degree of polymerization, as in oligonu-cleotide, oligopeptide, oligosaccharide. (*see also* mono-; oligomer; poly-)

oligomer A small polymer of complexity greater than that of a monomer but less, in common usage, than that of a dodecamer.

oligomerization (*see* aggregation)

oligonucleotide array (= chip)

oligonucleotide ligation assay (OLA) A method to identify a specific oligonucleotide sequence without use of radiochemicals, electrophoresis or centrifugation. An oligonucleotide biotinylated at its 5'-end and another with a reporter group (chromophore or fluorophore) at its 3'-end are constructed to hybridize to the sequence to be detected in a template DNA strand. A ligase is able to join the two oligonucleotides if they both match the template perfectly. The ligated oligonucleotide can be extracted from the mixture on to immobilized streptavidin and the presence of the reporter group detected.

• Nickerson, D.A., Kaiser, R., Lappin, S., et al. (1990) Proc Natl. Acad. Sci. U.S.A. **87**, 8923–8927

oligosaccharin One of a group of complex oligosaccharides shown to have a regulatory function in plants, i.e. growth regulation, organogenesis, defence against pathogens.

• Albersheim, P., Darvill, A., Augur, C., et al. (1992) Acc. Chem. Res. **25**, 77–83

O-linked carbohydrate The carbohydrate chains that are attached by a glycosidic bond to a protein through the hydroxy group of a serine or threonine residue. (*see also* N-linked carbohydrate)

omega (Ω)-loop An irregular secondary structure on the surface of many globular proteins that consists of 6–16 amino acid residues folded into a rigid, tightly packed loop in which the N- and C-terminal ends are located close together, thus forcing the sequence into an Ω-shaped feature.

OMH (= optimal matching hydrophobicity)

oncogene A gene present in normal cells which when altered can transform the normal cell into a malignant cell. (When emphasizing this relationship, the unaltered gene is called a proto-oncogene and the altered gene, an oncogene.) The alteration may result in overproduction of a gene product or faulty function. An oncogene is denoted by the prefix c if a cellular oncogene, e.g. c-*ras*, or by the prefix v if a viral oncogene, e.g. v-*ras*, which is presumed to have originated from a capture of the c-oncogene by the virus during an infection of a normal cell. Contrasted with a *thanatogene*, or *death gene*, which when activated leads to apoptosis. (*see also* anti-oncogene; oncoprotein)
• Marx, J. (1994) Science **266**, 1942–1944

oncoprotein The translation product of an oncogene; a regulator of cellular processes, which if the gene is mutated, results in uncontrolled growth (cancer). Oncoproteins may be embedded in the plasma membrane and serve as receptors, reside in the cytosol and be involved in signal transduction, often associated with a tyrosine kinase activity, or may be found in the nucleus and be directly involved in gene regulation. (*see also* oncogene)

one-by-one type mechanism (*see* zipper-type mechanism)

one-carbon pool The aggregate of freely interconvertible moieties that contain single carbons more reduced than carbon dioxide; i.e. metabolites of tetrahydrofolate, the methyl groups of choline, the hydroxymethyl of serine, etc.

one-electron carrier A compound that can accept or donate a single electron; e.g. a cytochrome.

one gene one enzyme hypothesis The proposal that arose from genetic studies on the effects on amino acid metabolism of mutations in *Neurospora*, that there is a one-to-one correspondence between enzymes and the genes that encode them.

one-hybrid system (*see* two-hybrid system)

oocyte A female gamete, a precursor of an ovum.

opal mutation A nonsense mutation that, due to the premature appearance in the mRNA of the terminator codon UGA, leads to the formation of a truncated, possibly non-functional, protein. (*see also* amber mutation; ocher mutation)

open-chain form The form of a sugar in which there has been no internal condensation to form a pyranose or furanose. (*see also* anomer; alpha-isomer; beta-isomer)

open reading frame (ORF) One of three possible reading frames in which an mRNA is potentially translated into protein. In analysis of a DNA sequence, an ORF is characterized by the sequence of nucleotides that when transcribed into mRNA results in a series of triplet codons that is not interrupted by a translation termination codon. (*see also* unassigned reading frame)

operator A locus on DNA that controls transcription when a repressor or activator becomes bound.

operon A group of genes, usually metabolically related, that can be controlled co-ordinately from a common regulatory sequence. (*see also* cistron)

opiate Morphine, pharmacologically active derivatives of it or compounds such as enkephalins and endorphins that similarly act at the same receptors in the brain.

opsonin The immunoglobulin components of complement attached to a foreign cell that sensitize it to phagocytosis.

optical rotation The change in the angle of transmitted plane-polarized light as it passes through a solution of a chiral sample, such that the change is proportional to the concentration of the chiral molecule and to the length of the light path through the solution. In *dextro*-rotation the change is clockwise when facing the light source; in *laevo*-rotation the change is counterclockwise. (*see also* circular dichroism; optical rotatory dispersion)

optical rotatory dispersion (ORD) The optical rotation of a material as a function of wavelength. (*see also* circular dichroism)

optimal matching hydrophobicity (OMH) A calculated value used to predict conformational similarities of proteins from their amino acid sequences. Numerical values for hydrophobicity, OMHs, are assigned to each amino acid residue to allow comparisons based upon features of the residues that are important to their structural function, but independent of actual amino acid identities. An OMH plot, OMH values against sequence number, permits visual comparisons of proteins. (*see also* hydropathy index; hydropathy plot)

optode A biosensor with a light-detecting sensor, e.g. a device that detects the lowering of pH due to the action of glucose oxidase at the end of a fibre optical light guide by the effect on a pH-sensitive fluorescent dye.

• Turner, A.P.F. (1992) Essays Biochem. **27**, 147–159

orbital perturbation theory (*see* entropy effect)

orbital steering (*see* entropy effect)

ORC (= origin recognition complex)

ORD (= optical rotatory dispersion)

ordered (*see* enzyme mechanism)

ORF (= open reading frame)

organelle Inclusion in the cell cytoplasm that can be membrane-limited (e.g. chloroplast, mitochondrion, endoplasmic reticulum, Golgi apparatus) or non-membrane-limited (e.g. cytoskeletal element, ribosome, nucleosome).

origin of replication A double-stranded DNA sequence that binds the proteins that will begin unwinding the helix, preparatory to initiation of replication.

origin recognition complex (ORC) An aggregate of nuclear proteins involved in initiation of replication and/or repression of transcription (silencing). The DNA regions that bind the ORC are examples of *anomalously replicating sequences* (*ARSs*), which promote plasmid replication in yeast. Other examples are *silencers*, which flank repressed genes;

the sequences that interact directly with an ORC are *ARS consensus sequences* (*ACSs*).

- Newlon, C.S. (1993) Science **262**, 1830–1831

orphon A displaced genetic element; a coding or non-coding sequence that usually occurs in a series of tandem repeats in a eukaryotic genome but appears in a location atypical of the species or strain.

orthologous Descriptive of homologous genes that have evolved divergently after speciation, e.g. cytochromes *c* of various species. (*see also* paralogous)

osazone The product of coupling the carbonyl carbon of a sugar with phenylhydrazine, its further oxidation and subsequent addition of a second phenylhydrazine to form a crystalline compound.

osmolarity The concentration of non-permeable solutes that contribute to osmotic pressure; *iso-osmotic* or *isotonic* if equal to that of a cell, *hypotonic* if lower and *hypertonic* if higher.

osmotic pressure The pressure developed by a solution of a non-permeable solute isolated by a membrane from a surrounding solution.

osteo- A prefix that indicates bone, e.g. osteocyte.

Ouchterlony plate (*see* immunodiffusion)

outer mitochondrial membrane (*see* mitochondrion)

outer nuclear membrane (*see* nuclear envelope)

overhang The extension of one polynucleotide strand beyond the terminus of its complementary strand.

Overhauser effect (= nuclear Overhauser effect; *see* nuclear Overhauser and exchange spectroscopy)

overlapping gene A rare phenomenon found in small viral genomes whereby a single mRNA can be translated in two different reading frames to produce two different proteins.

ovum A female gamete, which can be fertilized by a spermatozoon.

oxidation–reduction potential (= redox potential)

oxidative decarboxylation Removal of carbon dioxide from a carboxylic acid that is facilitated by oxidation of the α-carbon (*α-oxidative decarboxylation*, e.g. the pyruvate dehydrogenase reaction) or the β-carbon (*β-oxidative decarboxylation*, e.g. the isocitrate dehydrogenase reaction).

oxidative phosphorylation The reactions that synthesize the phosphoanhydride bond of ATP by the coupling of that energetically unfavourable reaction with the spontaneous oxidation of metabolites. (*see also* chemiosmotic theory; P:O ratio)

oxidative stress The cumulative damage due to the less than 100% effectiveness of antioxidants in prevention of free radical reactions such as lipid peroxidation.

oxidoreductase One of a class of enzymes that oxidize one substrate as they reduce another, e.g. a dehydrogenase, an oxidase, a peroxidase.

oxygen dissociation curve A characterization of an oxygen-binding protein, such as haemoglobin; a plot of percentage saturation against partial pressure of oxygen.

P

p In addition to use as the metric notation of pico (10^{-9}), it is commonly used to indicate the logarithm of a reciprocal, e.g. $pK_a = \log(1/K_a)$. When seen as a prefix followed by a number, p indicates molecular mass determined from a Laemmli gel and characterizes a protein whose identity is otherwise unknown, e.g. p53, a protein of mass 53 000 Da. It is also used as a prefix to an alphanumeric designation of a plasmid, e.g. pVA50. In nucleic acid chemistry it denotes an orthophosphate in a 5'- or 3'-phosphomonoester bond (e.g. pA or Gp), a phosphodiester bond (e.g. CpG), or a phosphate anhydride bond (e.g. ppA). In genetics it is a designation of a location in the short arm of a chromosome. (*see also* centromere)

packaging cell In biotechnology, a eukaryotic cell that is used to produce a modified virus.

PAD (= pulsed amperimetric detection)

PAGE Polyacrylamide-gel electrophoresis. (*see* electrophoresis; native gel; sodium dodecyl sulphate/polyacrylamide-gel electrophoresis)

palindrome A concept borrowed from linguistics; a segment of a double-stranded polynucleotide such that the order of bases, read 5'-to-3' in one strand, is the same as that in the complementary antiparallel strand, read 5'-to-3'; a common structural feature of DNA sequences that constitute specific protein-binding sites.

pan editing (*see* guide RNA)

paracellular (*see* tight junction)

paracrine Descriptive of a secretion that acts on cells adjacent to the site of its secretion. (*see also* autocrine; endocrine; exocrine; telecrine function)

paradox of Levinthal The discrepancy between the enormous number of conformations a polypeptide may assume and the rapidity with which it normally achieves its native conformation. It excludes the possibility that each conformation is sampled and only the thermodynamically most stable form persists. (*see also* framework model)

parallel synthesis (*see* small molecule library)

paralogous Descriptive of homologous genes that have duplicated and evolved divergently, e.g. those encoding trypsin, chymotrypsin, elastase and thrombin, which are all serine proteinases and occur together within the same species. (*see also* orthologous)

paramagnetic Descriptive of a chemical species that can interact with a magnetic field, usually a transition metal ion or an organic compound with an unpaired electron. (*see also* diamagnetic)

paranemic (*see* plectonemic)

paratope The antigen-binding site of an antibody. (*see also* epitope)

parenchyma The functional tissue of an organ, as contrasted with the stroma.

parenteral nutrition Provision of food other than orally, usually by intravenous infusion.

PARF (= polymorphic amplifiable restriction endonuclease fragment)

partition chromatography A technique for separation of molecules based on their differing ratios of solubility in two immiscible solvent phases. A stationary phase is coated or impregnated with one solvent phase and the mixture to be separated is passed over it in the mobile phase.

partition coefficient The ratio of solubility of a substance in two immiscible solvents, e.g. in oil and water.

partner One of two different helix–loop–helix proteins that together form a complex that can bind to DNA and regulate its expression.
• Barinaga, M. (1991) Science **251**, 1176–1177

passive diffusion The transport of compounds across a membrane that is unmediated by any mechanism and is independent of energy sources, and therefore occurs at a rate that is determined by the area of the membrane, the concentration difference across it and the solubility of the compound in the membrane. (*see also* active transport; facilitated diffusion)

PAS stain (= periodic acid–Schiff stain)

Pasteur effect The inhibition by oxygen of glucose consumption in a tissue or microbiological preparation. (*see also* Crabtree effect)

patch-clamp technique A method for measurement of conductance of cell membranes by clamping a tiny buffer-filled pipette tip to a cell surface, so that the pipette serves as an electrode for the measurements.

patched circle PCR A method for site-directed deletion of a large-scale fragment from a cloning vector, e.g. for obtaining cDNA from a plasmid without vector sequences. Two PCR primers pair to the 5'- and 3'-flanking sequences adjacent to and extending away from the unwanted sequence, and the DNA is amplified by PCR. Another oligonucleotide is added to bind to both ends of the amplified linear cDNA and patches them together to mimic closed circular DNA, which is then used to transform *E. coli.*
• Song, C. and Yang, K. (1991) Biotech Forum Eur. **8**, 130–132.

patching The migration of receptors to certain areas on the surface of cells when exposed to ligands, especially the interaction of an antibody with its receptors on the surface of lymphoid cells; also called *capping*.

pathway (= metabolic pathway)

paucidisperse (*see* monodisperse)

Pauly reaction A colorimetric reaction for identification of imidazole compounds that involves reaction with diazobenzenesulphonate.

PCD Programmed cell death. (= apoptosis)

PCR (= polymerase chain reaction)

PCR-based differential scanning A method for amplification of a heterogeneous population of cDNAs preparatory to creation of a library. Total mRNA is copied first using an oligo(dT) primer to obtain one copy, then that primer is removed and the polynucleotides are tailed with oligo(dC) so that the second strand can be synthesized using an oligo(dG) primer. This second strand can then be amplified by PCR using oligo(dT) and oligo(dG) primers.
• Brunet, J.-F., Shapiro, E., Foster, S.A., et al. (1991) Science **252**, 856–859

PD-MS Plasma-desorption mass spectrometry. (*see* mass spectrometry)

pellet The sedimented portion that accumulates during centrifugation. (*see also* supernatant fluid)

pentacovalent intermediate An intermediate or transition state in phosphate reactions, in which an incoming oxygen atom displaces a leaving oxygen in a concerted reaction and forms a *trigonal bipyramid* in which the phosphorus atom is at the centre, three oxygens are co-planar around it and the entering and leaving oxygens are normal to the plane on opposite sides of the phosphorus atom.

pentose nucleic acid Obsolete name for RNA.

pentose phosphate pathway The enzymic reactions that oxidatively convert glucose 6-phosphate into ribulose 5-phosphate and then to intermediates of the glycolytic pathway. The two oxidative reactions are the major source of NADPH in many cells. Also known as the Dickens–Warburg pathway, the hexose monophosphate shunt, the pentose shunt and the phosphogluconate oxidative pathway.

pentose shunt (= pentose phosphate pathway)

penultimate Descriptive of a position in a linear sequence, e.g. of a polypeptide, adjacent to one of the terminal positions.

peptidase An enzyme that cleaves the peptide bonds of proteins and peptides. (*see also* endopeptidase; exopeptidase).

peptide A compound formed by incomplete hydrolysis of a protein, or by elimination of the elements of water from between the α-carboxy and the α-amino groups of α-amino acids to form a linear polymer.

peptide bond The amide bond formed by condensation of the α-carboxy group of one amino acid with the α-amino group of another; the bond that joins together the amino acid residues that comprise a protein, peptide or polypeptide.

peptide nucleic acid (PNA) A synthetic oligonucleotide analogue in which the charged poly(deoxy)ribosylphosphate backbone is replaced by a neutral peptide chain, upon which the nucleoside residues are strung.
• Bruice, T.C., Dempcy, R.O. and Browne, K.A. (1995) J. Am. Chem. Soc. **117**, 6140–6142

peptide site The part of a ribosome that binds the growing peptidyl-tRNA before the peptidyl group is transferred to the next amino acyl residue which is held at the amino acyl site as its tRNA ester.

peptide Velcro A description of helix–helix interactions in which side chains of two parallel, adjacent helices closely interact due to complementary shapes and electrostatic interactions, e.g. the leucine zipper motif of the fos and jun heterodimers.
• O'Shea, E.K., Lumb, K.J. and Kim, P.S. (1994) Curr. Biol. **3**, 658–667

peptidoglycan A polymer of bacterial cell walls that exhibits considerable species variation in structure. It is generally composed of a heterodisaccharide attached to a peptide; the carbohydrate moieties of neighbouring subunits are cross-linked, and the peptides of neighbouring subunits are also cross-linked.

peptidyl Descriptive of a peptide as it is attached by its α-carboxy group, e.g. a peptidyl-tRNA.

peptoid A proteinase-resistant peptide analogue, e.g. a polymer in which the amino acid side chains are shifted from the α-carbons to the amide nitrogens.

PER Pre-edited region. (*see* guide RNA)

perfusion An *in vitro* or *in situ* technique to circulate a fluid through the blood vessels of an organ. (*see also* perifusion)

perfusion chromatography A variant of high-pressure liquid chromatography in which particles as the stationary phase have pores large enough [6000–8000 Å (600–800 nm)] to permit solvent flow-through and thus allow much faster flow rates, decreased diffusional spreading and enhanced access to the full surface of the stationary phase.
• Regnier, F.E. (1991) Nature (London) **350**, 634–636

perifusion An *in vitro* technique to circulate fluid around an isolated organ or tissue held in a specially designed chamber. (*see also* perfusion)

perinuclear space The space between the nuclear inner and outer membranes.

periodic acid–Schiff (PAS) stain A reagent for staining of carbohydrate in polyacrylamide gels or histological preparations; periodic acid first oxidizes vicinal hydroxy groups to generate aldehydes that can then condense with fuchsin sulphurous acid (Schiff's reagent) to form an intensely coloured Schiff base.

peripheral nervous system (PNS) A division of the nervous system of higher animals that consists of the neurons that extend from the spinal cord to skeletal muscles, glands, etc. and the neurons that return from these structures to the spinal cord. (*see also* central nervous system)

peripheral protein (extrinsic protein) A protein loosely associated with a membrane. (*see also* integral protein)

peripheral tissue In any particular discussion, tissues that are not of central concern; e.g. in urea synthesis, tissues other than liver; in digestion, tissues other than the gastrointestinal tract.

periplasmic space The space between the cell membrane and the outer cell wall of some non-animal and prokaryotic cells.

permeability coefficient A quantitative measure of the rate at which a molecule can cross a membrane such as a lipid bilayer; expressed in units of cm/s and equal to the diffusion coefficient divided by the width of the membrane.

permease An active transport mechanism.

permissive host A cell in which a virus is able to replicate. (*see also* non-permissive host)

permissive temperature A temperature at which a temperature-sensitive mutant will grow.

peroxidase A haem enzyme that abstracts two hydrogens from one substrate to reduce hydrogen peroxide, its other substrate, to water.

peroxisome (microbody) A vesicle containing various oxidizing enzymes (which generate H_2O_2) and catalase (which degrades H_2O_2).

Perrin plot A graphical method to determine limiting anisotropy or limiting polarization; a display of the reciprocal of polarization against the ratio of absolute temperature to viscosity. The anisotropy or polarization is measured at increasing viscosities (effected by the addition of a solute such as sucrose or glycerol) or decreasing temperatures and extrapolated to absolute zero or infinite viscosity.

PERT (= phenol-emulsion reassociation technique)

PEST hypothesis The proposal that polypeptide sequences rich in the amino acid residues proline (P), glutamate (E), serine (S) or threonine (T) render a cellular protein susceptible to rapid turnover.

• Rivett, A.J. (1990) Essays Biochem. **25**, 39–81

petite mutation Mutation observed in a strain of yeast that is deficient in mitochondrial protein synthesis. The mutants were instrumental in the discovery of mitochondrial DNA and mitochondrial protein synthesis.

PFGE (= pulsed-field gel electrophoresis)

PFU (= plaque-forming unit)

pH A measure of acidity of aqueous solutions; the negative of the common logarithm of the hydrogen ion concentration, $-\log[H^+]$. Values below 7 are acidic; values above 7 are alkaline.

phage (= bacteriophage)

phage display A method for incorporation, into the DNA that encodes a phage coat protein, of a polypeptide, where it is available for a selection procedure; especially for production and selection of antibody (even anti-self antibody) fragments directed against a specific antigen. A library of V_H or V_L genes is produced by PCR using primers directed to constant regions that flank the variable region, cloned, and inserted into the gene for a phage coat protein that, when expressed, is available for cycles of growth and selection by adsorption on to an antigen-coated plate or beads.

• Makowski, L. (1994) Curr. Opin. Struct. Biol. **4**, 225–230; Tramontano, A. and Sollazzo, M. (1995) Curr. Opin. Biotechnol. **6**, 73–80; O'Neil, K.T. and Hoess, R.H. (1995) Curr. Opin. Struct. Biol. **5**, 443–449

phagemid A multifunctional cloning vector; a chimaera of a plasmid and a bacteriophage that incorporates and consolidates many useful features of both.

• Watson, J.D., Gilman, M., Witkowski, J. and Zoller, M. (1992) Recombinant DNA, 2nd edn., p. 119, Scientific American Books, New York

phagocytosis (*see* endocytosis)

phagolysosome A vesicle formed by fusion of a phagosome with a lysosome.

phagosome A vesicle that forms in a cell as a result of phagocytosis and that contains the engulfed material.

pharming (= molecular pharming)

phase problem In X-ray crystallography, the gap between the requirements for reconstruction of a model – for each reflection (defined by the angle of its diffraction), the amplitude (related to intensity) and phase (a measure of the alignment of the wave function with that of other reflections) – and the experimentally accessible quantities, the intensities of each reflection. (*see also* diffraction pattern; Fourier transformation)

phase transition A melting-like phenomenon that occurs when lipid bilayers are warmed, usually below the naturally occurring temperature of a plasma membrane; due to thermal disruption of the regular stacking of fatty acid side chains. (*see also* transition temperature)

phenol-emulsion reassociation technique (PERT) A method to accelerate several thousand-fold the rate of DNA reassociation. An aqueous solution of single-stranded DNA with a chaotropic salt is shaken with phenol. When the resulting emulsion separates into two phases, the polynucleotide is found to have rapidly reassociated into double-stranded DNA. (*see also* FPERT)

phenotype (*see* genotype)

pheromone A chemical signal that travels through air or water from one organism to another.

***phi*-angle (ø angle)** A characteristic of a peptide bond; the angle formed at the nitrogen atom by the bonds to the α-carbon and to the carbonyl carbon of the neighbouring residue. (*see also psi*-angle; Ramachandran plot)

pH-jump (*see* relaxation)

phorbol ester A tetracyclic diterpene tumour-promoting substance originally derived from croton oil.

phosphagen A compound that acts as a storage form of high-energy phosphate, e.g. phosphocreatine, phosphoarginine.

phosphate ester A structure formed by the removal of the elements of water from between a phosphoric acid and an alcohol, e.g. AMP, glucose 6-phosphate.

phosphate gripper A basic oligopeptide sequence of many enzymes that protects their phosphate-containing substrate or transition state from interaction with water.

• Knowles, J.R. (1991) Nature (London) **350**, 121–124

phosphate transfer potential (*see* group transfer potential)

phosphoacylglycerol (phosphoglyceride) Phosphatidic acid or a derivative, e.g. phosphatidylcholine, phosphatidylserine.

phosphoanhydride A structure formed by the removal of the elements of water from between two phosphoric acids, e.g. pyrophosphate, ADP.

phosphodiester A diester of phosphoric acid; in nucleotide chemistry may be an internal diester, e.g. cyclic AMP, or an internucleotide bond as in RNA and DNA.

phosphogluconate oxidative pathway (= pentose phosphate pathway)

phosphoglyceride (= phosphoacylglycerol)

phosphoinositide cascade A series of events that is initiated extracellularly and leads via activation of phospholipase C to the liberation from membrane phospholipids of inositol 1,4,5-trisphosphate and diacylglycerol, which act intracellularly to increase cytosolic calcium and to activate protein kinase C respectively.

phospholipid A covalent association of phosphoric acid and lipid, e.g. a phosphoacylglycerol, sphingomyelin.

phospholipid-transfer protein (exchange protein) A lipoprotein of the intracellular membranes of eukaryotic cells that can exchange its phospholipid or sterol for membrane components, and can shuttle lipids between these membranes, e.g. between the endoplasmic reticulum and Golgi apparatus.

• Rothman, J.E. (1990) Nature (London) **347**, 519–520

phosphorescence A phenomenon similar to fluorescence in which there is a long lag between excitation and emission. (*see also* triplet state)

phosphoroclastic split A cleavage in which the elements of phosphate are inserted across a carbon–carbon bond, e.g. the phosphoroclastic reaction of pyruvate, in which pyruvate and orthophosphate are oxidized by a flavoprotein to acetylphosphate and CO_2.

phosphorolysis Analogous to hydrolysis, the cleavage of a covalent bond by insertion across it of the elements of phosphoric acid, e.g. the glycogen phosphorylase reaction.

phosphorothioate A phosphate analogue, especially a nucleotide analogue, in which a sulphur atom replaces an oxygen in one of the phosphate groups.

phosphorylation potential A quantitative measure of the energy status of a cell: $[ATP]/[ADP][P_i]$. (*see also* adenylate energy charge)

photo-activated repair A mechanism for repair of DNA by reduction of thymine dimers by DNA photolyase; contrasted with *dark repair*, an ATP-dependent process that involves excision of the dimers, resynthesis of the sequence by a polymerase and rejoining of the ends by a ligase.

photoaffinity labelling A variant of affinity labelling in which the covalent attachment of a ligand to its site is effected by illumination of a light-sensitive chemical group on the ligand, often an azido group.

photocycle The series of transformations undergone by rhodopsin from its capture of radiant energy, i.e. *cis–trans* isomerization of the all-*trans* form and regeneration of all-*trans*-rhodopsin.

photoisomerization The light-induced rearrangement of one of the double bonds of all-*trans*-retinal.

photoluminescence (*see* chemiluminescence)

photon The fundamental particle of light.

photorespiration The reaction of ribulose bisphosphate carboxylase in which molecular oxygen replaces carbon dioxide as a substrate, to form 3-phosphoglycerate and phosphoglycolate rather than two 3-phosphoglycerates.

photosynthesis The light-dependent chlorophyll-catalysed biosynthesis by green plants of carbohydrate and oxygen from carbon dioxide and water. (*see also* Calvin cycle; C_3–C_4 photosynthesis; C_3 photosynthesis; C_4 photosynthesis; Hill reaction)

photosystem I The light-driven reactions of photosynthesis that absorb at 700 nm or below and result in generation of NADPH. (*see also* photosystem II)

photosystem II The light-driven reactions of photosynthesis that absorb at 680 nm or below and result in oxidation of water to molecular oxygen. (*see also* photosystem I; red drop)

• Rogner, M., Boekema, E.J. and Barber, J. (1996) Trends Biochem. Sci. **21**, 44–49

phototroph An organism that derives its energy by trapping radiant energy from the sun or other light source. (*see also* chemotroph)

pH stat A device that automatically delivers and records against time the volume of acid or alkali necessary to maintain a preset pH during a reaction that consumes or generates acid.

phycobilin (*see* bilin)

phycofluor A cytological probe; a synthetic conjugate of an intensely fluorescent phycobiliprotein with a molecule like an antibody that gives it specificity.

phylogenetic tree (evolutionary tree) A reconstruction of the evolutionary relationship of contemporary species that shows their divergence from common ancestors and the branch points at which different species separated. A common ancestor is presented at the base of a central stem, and the tree ramifies as it ascends, often with distance from the base roughly related to evolutionary time. The relationships it displays are usually inferred from the similarities of amino acid sequences of contemporary proteins or of the base sequences of contemporary nucleic acids.

physical map A goal of the human genome project; the characterization of each chromosome by landmarks, sequence-tagged sites and contigs, between which restriction maps may give finer detail.

• Olson, M., Hood, L., Cantor, C. and Botstein, D. (1989) Science **245**, 1434–1435

P_i Inorganic orthophosphate, i.e. PO_4^{3-} and protonated forms.

picket fence porphyrin compound A model oxygen-binding porphyrin in which bulky side chains on one face of the ring permit binding of molecular oxygen while preventing oxidation of Fe^{2+}, which is co-ordinated on the other face of the ring by an organic base.

pi electron–pi electron (π electron–π electron) interaction A form of non-covalent interaction seen in proteins; the planes of two aromatic side chains overlap so that the π orbitals can interact.

piezophile (*see* extremophile)

PIG Phosphatidylinositolglycan. (*see* glypiated)

pigylated (= glypiated)

ping pong (*see* enzyme mechanism)

pinocytosis (*see* endocytosis)

pitch A feature of a helix, the distance parallel to the axis that corresponds to one turn of $360°$, e.g. 5.4 Å (0.54 nm) for an α-helix. (*see also* repeat; rise)

pK_a The negative of the common logarithm of the dissociation constant of an acid; i.e. $-\log K_a$, where $K_a = [H^+][A^-]/[HA]$ for $[HA] = [H^+] + [A^-]$.

PKC (= protein kinase C).

PKS Polyketide synthase. (*see* polyketide)

plaque (*see* plaque-forming unit)

plaque assay A technique for quantification of infectious phage particles by counting plaque-forming units.

plaque-forming unit (PFU) A measure of viable bacteriophage particles, determined by plating a known volume of a solution on to a bacterial lawn and subsequently counting the areas of bacterial lysis (*plaques*).

plaque hybridization (*see* colony hybridization)

plaque purification A technique to select a bacterial strain with a desired genetic trait; a mixture of cells is streaked on to a solid support and the resultant colonies are screened (e.g. for antibiotic resistance or a DNA sequence complementary to a specific probe), and the selected colony is then restreaked for further rounds of purification.

plasma In biology, the liquid phase of whole blood. (*see also* serum)

plasma cell A differentiated B-cell that secretes antibodies.

plasma-desorption mass spectrometry (*see* mass spectrometry)

plasmalemma (= plasma membrane)

plasma membrane The lipid bilayer and associated proteins and other molecules that surround a cell.

plasmid A self-replicating extra-chromosomal element, usually a small segment of duplex DNA that occurs in some bacteria; used as a vector for the introduction of new genes into bacteria.

plasmolysis The dehydration of a cell caused by exposure to a hypertonic solution.

plate A surface of solid growth media, often in a Petri dish, on which a bacterial culture can be grown; also the action of streaking a culture on the media.

platelet A small membrane-limited fragment of cytoplasm derived from blebs from a megakaryocyte; platelets aggregate to form a plug during early stages of haemostasis.

pleated sheet (= beta-pleated sheet)

plectonemic Descriptive of the interaction of two DNA strands, either single- or double-stranded, in which an oligonucleotide of one strand is twisted into a plait of helices, one around the other, and requires the nicking of one strand to form or dissociate such an interaction. When one of the DNA strands is circular, the structure, a ring pierced by the linear strand, is called a *hemicatenane*.

Plectonemic is contrasted with *paranemic*, descriptive of an interaction between DNA strands that is formed and dissociated without the breaking of covalent bonds and therefore constitutes a weaker interaction than a plectonemic joint. Plectonemic and paranemic are often used to describe joints formed during recombination events. A related folding is *toroidal*, in which a plait of helices is circularized and forms a ring, or *torus*.

pleiotropic mutation A mutation that affects expression of several characteristics, e.g. the effect of mutation of a promoter on structural genes regulated by it.

pleiotropism A characteristic of action of some hormones; expression of several activities, e.g. the action of insulin both to promote glucose uptake by skeletal muscle cells and to promote protein synthesis.

pleuropneumonia-like organism (= mycoplasma)

plus and minus method A variant of the Sanger method of DNA sequencing. A primer is extended by a polymerase to generate a population of newly synthesized deoxyribonucleotides of assorted lengths; the unused dNTPs are removed, and polymerization continues in four pairs of plus and minus reaction mixtures; the minus mixtures have three NTPs and the plus mixtures have only one. After a second polymerization, the mixtures are fractionated by gel electrophoresis, and each plus and minus pair is compared to indicate the length of the new polydeoxyribonucleotide (by the mobilities of the bands) and the position at which polymerization had terminated as a result of the absence of the missing dNTP.

• Wu, R. (1994) Trends Biochem. Sci. **19**, 429–433

plus-strand The coding DNA strand during transcription, as opposed to the non-coding *minus-strand.*

PMN (= polymorphonuclear leucocyte)

PNA (= peptide nucleic acid); also pentose nucleic acid; an obsolete name for RNA.

PNS (= peripheral nervous system; or = positive and negative selection)

pocket An invagination of the surface of a protein where it can bind a ligand. (*see also* cleft)

poikilothermic Descriptive of an organism that cannot maintain a fixed internal temperature. (*see also* homoiothermic)

point mutation A change in a single nucleotide residue in a nucleic acid, which frequently causes a single amino acid residue replacement in the protein product of the gene.

polarization A measure of the mobility of a fluorophore: $P = (I_\| - I_\perp)/(I_\| + I_\perp)$, where I is the intensity of emission and $\|$ and $\perp$ indicate polarization that is parallel and perpendicular respectively to the exciting light. A *mobile fluorophore* is able to reorient itself within its fluorescence lifetime and therefore emits unpolarized light; an *immobile fluorophore* does not reorient and hence emits light polarized in the plane of excited light. (*see also* anisotropy)

pole of a cell (*see* centriole)

poly- A prefix that indicates many; a high degree of polymerization, as in polynucleotide, polypeptide. (*see also* mono-; oligo-)

polyamine A compound with more than one amino group that often contains short chains of carbon atoms separated by a nitrogen atom and forms a secondary amine; associated in the cell with nucleic acids; e.g. spermine, spermidine, putrescine.

polyampholyte A large molecule or small polymer that contains many acidic and basic groups. (*see also* isoelectric focusing; polyelectrolyte)

poly(A)⁺ RNA Polyadenylated RNA; usually synonymous with mature eukaryotic mRNA; contrasted with *poly(A)⁻ RNA*, which is non-polyadenylated RNA.

poly(A)⁻ RNA [*see* poly(A)⁺ RNA]

poly(A) tail The post-transcriptional modification of the 3′-end of eukaryotic mRNA; a sequence of 20–200 adenylates in 5′-to-3′ linkage.

polycistronic mRNA A single mRNA that contains the information necessary for the production of more than one polypeptide; characteristic of some prokaryotic mRNAs. (*see also* cistron; monocistronic mRNA; operon)

polyclonal antibody A heterogeneous immunoglobulin preparation that contains antibodies directed against one or more determinants on an antigen; the product of daughters of several progenitor cells that have been programmed for immunoglobulin synthesis and secretion. (*see also* monoclonal antibody)

polydisperse (*see* monodisperse)

polyelectrolyte A compound or polymer with many charges of the same kind, e.g. polyglutamate, polylysine.

polyene A multiply unsaturated compound, especially one that has strong visible absorbance bands and is capable of transducing light, e.g. chlorophyll, retinal.

polyketide A compound derived from biosynthetic acetyl and/or propionyl units and other building blocks, in which carbonyl carbons condense with α-carbons analogously to the synthesis of fatty acids. They are often fungal metabolites such as macrolides, which are formed by *polyketide synthases* (*PKSs*), modular multienzyme complexes encoded as clusters of open reading frames (ORFs) spanning more than 100 kbp of the microbial genome, with each ORF responsible for a different enzyme activity. The cluster for any given PKS recruits the ORFs for those transferases that can form the carbon skeleton and the reductases and dehydratases that transform it into the particular polyketide product.

• Schwecke, T., Aparicio, J.F., Molnar, J., et al. (1995) Proc. Natl. Acad. Sci. U.S.A. **92**, 7839–7843

polyketide synthase (*see* polyketide)

polylinker A short DNA sequence that contains several restriction sites; intended to be placed into a vector so that the sites may subsequently be used for insertion of other DNA sequences.

polymerase chain reaction (PCR) A technique to amplify a specific region of double-stranded DNA. An excess of two *amplimers*, oligonucleotide primers complementary to two sequences that flank the region to be amplified, are annealed to denatured DNA and subsequently elongated, usually by a heat-stable DNA polymerase from *Thermus aquaticus* (*Taq* polymerase). Each cycle involves heating to denature double-stranded DNA and cooling to allow annealing of excess primer to template and elongation of the primers by the *Taq* polymerase; the number of *amplicons*, i.e. the target sequence fragments between flanking primers, doubles with each cycle.

polymorphic amplifiable restriction endonuclease fragment (PARF) A natural polymorphism detectable with probes constructed to a representation of the affected parent DNA.

• Lisitsyn, N., Lisitsyn, N. and Wigler, M. (1993) Science **259**, 946–951

polymorphism Appearance in several forms, e.g. of a protein that is subject to tissue-specific post-translational modification, or of mRNA that is processed through alternative splice sites; also the appearance in a population of more than one structural gene at a single genetic locus.

polymorphonuclear leucocyte (PMN) One category of blood cell; a neutrophil.

polyol A simple sugar whose carbonyl function has been reduced, e.g. sorbitol, mannitol.

polyprotein A primary product of protein synthesis, a single polypeptide chain that will eventually be cleaved into several separately functional proteins; e.g. the precursor produced by retroviruses that gives rise to several proteins, among them a coat protein, a polymerase and an endopeptidase.

polysome A structure that is functional in protein synthesis, formed by several ribosomes attached to a single mRNA molecule.

polytene chromosome An amplified form of an insect chromosome in which the chromatin is packed in irregularly spaced transverse bands.

polytopic Descriptive of a protein that spans a membrane at least once and has domains that project into both cytoplasmic and extracellular spaces.

pool The aggregate of metabolites that are in a functional equilibrium in a cell or an organism, e.g. cytosolic hexose phosphates, mitochondrial tricarboxylic acid cycle intermediates, liver glycogen. (*see also* compartment)

P:O ratio The ratio of phosphate incorporated into ATP to oxygen atoms reduced to water; a measure of the efficiency of coupling of phosphorylation to oxidation. Based on the now-obsolete chemical coupling theory of oxidative phosphorylation, the passage of electrons down the electron transport pathway results at several points in the chemical synthesis of ATP. The number of phosphates that are fixed into ATP for every two electrons that pass from a substrate to reduce each oxygen atom of molecular O_2 (the P:O ratio) corresponds to the number of such points and should be an integer value. Experimentally, it was found to be between 2 and 3 when NADH was the donor of electrons and between 1 and 2 when succinate was the donor, and the theoretical integer values were set to 3 and 2 respectively. The chemiosmotic theory similarly predicts $H^+:O$ and $H^+:ATP$ ratios. Experimentally these appear to be 10 and 4 respectively when NADH is the substrate, equivalent to a P:O ratio of 2.5, and 6 and 4 respectively for FAD-linked substrates (e.g. succinate), equivalent to a P:O ratio of 1.5.

porin A bacterial membrane channel that allows free transport of small hydrophilic substances.

• Nikaido, H. (1994) J. Biol. Chem. **269**, 3905–3908

porter A protein or group of proteins in a membrane that is responsible for facilitated transport.

positional candidate approach A strategy for identification and cloning of a new gene; a combination of *functional cloning*, in which a property of the protein product (e.g. immunogenicity or liganding) is used to scan a cDNA library, and *positional cloning*, a more laborious approach that relies on chromosome walking from a nearby known sequence identified by genetic linkage.

• Ballabio, A. (1993) Nature Genet. **3**, 277–279

positional cloning (reverse genetics) The search for a gene (e.g. that for cystic fibrosis) with no knowledge of the gene product by locating it on the basis of linkage analysis with other genetic traits or individual polymorphisms in non-coding regions, followed by chromosome walking to identify candidate open reading frames that can be tested by expression in appropriate tissues or by zoo blotting, according to knowledge of the phenotype. Alternatively, the search for the gene for a specific protein from knowledge of its amino acid sequence (functional cloning), such that the sequence of part of the gene is deduced from the amino acid sequence and a labelled oligonucleotide probe complementary to it is synthesized. The probe allows one to screen appropriate DNA libraries for the gene. (*see also* positional candidate approach)

position effect The dependence of trans-gene expression on its site of incorporation into the host genome.

positive and negative selection (PNS) A technique to target insertion of a gene at a specific locus in a cell's genome. A plasmid is constructed in which a gene that allows positive selection (e.g. neomycin resistance) is inserted within the gene targeted for disruption, and a gene for negative selection (e.g. herpes simplex virus thymidine kinase, which is lethally sensitive to gancyclovir) is inserted outside the targeted gene sequence. When recombination is random the entire plasmid (including the negative selection gene) is incorporated into the cell's genome, but when recombination is with a homologous region the targeted gene is replaced only by the homologous region of the plasmid, including the gene for positive selection but excluding the gene for negative selection. (*see also* homologous replacement)

positive control The activation of transcription in a bacterium in response to its metabolic state, e.g. synthesis of several enzymes in response to glucose deprivation. (*see also* glucose effect)

• Adhya, S. and Garges, S. (1990) J. Biol. Chem. **265**, 10797–10800

positive co-operativity A form of allosteric behaviour in multimeric enzymes in which binding of substrate to one subunit increases the affinity of the other subunits for the substrate. (*see also* negative co-operativity)

positive energy balance The dietary situation in which caloric dietary intake is greater than energy expenditure, resulting in growth and possibly obesity. (*see also* negative energy balance)

positive-inside rule The generalization that, for proteins that span a membrane, the region that extends into the cytoplasmic space is positively charged.

post-labelling A method for detection of DNA modification. After exposure of DNA to a xenobiotic, it is digested with nucleases and labelled with ^{32}P; the resulting mixture is then separated by any of several conventional techniques to detect non-natural labelled fragments.

• Keith, G. and Dirheimer, G. (1995) Curr. Opin. Biotechnol. **6**, 3–11

post-synaptic Descriptive of a nerve cell that receives its signal from another nerve cell across a synapse. (*see also* pre-synaptic)

post-transcriptional modification The processing of RNA subsequent to its synthesis, which may include cleavage of phosphodiester bonds and modification of bases. For eukaryotic mRNA it includes capping of the 5′-end, polyadenylation of the 3′-end and splicing of introns.

post-translational modification The processing of a protein subsequent to its synthesis; includes phosphorylation, glycosylation, acetylation, limited proteolysis, attachment of prosthetic groups and cross-linking. (*see also* co-translational modification)

potentiometric titration The use of a potentiometer, especially a pH meter, to follow the titration of an oxidizable group with a reductant (or vice versa) to generate a graph of reduction potential against percentage reduction, or the titration of an acid with a base (or vice versa) to generate a graph of pH against amount of titrant.

potocytosis (*see* caveolae)

potonuon (*see* junk DNA)

Potter–Elvehjem homogenizer A device used to disrupt tissues. A cylindrical glass or hard polymer pestle rotates in a close-fitting tube and a suspension of the tissue particles is subjected to shearing forces as the pestle moves up and down and presses the suspension through the space between the rotating pestle and the tube.

PP$_i$ Pyrophosphate; $P_2O_7^{4-}$ and protonated forms.

prebiotic Descriptive of the synthetic reactions that occurred in geological time before the advent of life.

precipitin reaction The cross-linking of antigens through bivalent antibodies that creates an insoluble three-dimensional matrix.

precursor An antecedent metabolite whose structure is modified by one or more enzymic reactions of a metabolic pathway.

precursor protein (*see* protein splicing)

pre-edited region (*see* guide RNA)

pre-incubation The equilibration of some components before initiation of a reaction (the incubation) by addition of an essential substrate, cofactor or enzyme.

pre-initiation complex In replication, transcription or translation, a complex that is potentially composed of template, enzymes and initiation factors but that is incomplete for optimal function due to the absence of one or more elements. (*see also* initiation complex)

pre-mRNA (heterogeneous nuclear mRNA) The RNA that is the direct transcript from DNA and therefore includes introns before it is processed into mature mRNA.

prenyl A chemical group composed of multiples of the branched, unsaturated, 5-carbon isoprene unit, e.g. geranyl (10 carbons), farnesyl (15 carbons), geranylgeranyl (20 carbons). It is sometimes attached to the thiol group of an O-methylated C-terminal cysteine residue that forms part of a *CAAX box*, where C is the cysteine, A is any amino acid with an aliphatic side chain and X is any amino acid; a prenylpyrophosphate donates the prenyl group to the -CAAX sequence, AAX is proteolytically cleaved and the exposed carboxy group of the derivatized cysteine residue is enzymically methylated by *S*-adenosylmethionine. Prenylated proteins serve in the control of cellular function. Vitamins A and K, and dolichol pyrophosphate, carry reduced prenyl groups.

• Casey, P.J. and Seabra, M.C. (1996) J. Biol. Chem. **271**, 5289–5292

preparative *in situ* hybridization (prep-ISH) A coincidence cloning technique for isolation of DNA fragments that are characteristic of a small, well characterized portion of a chromosome. Restriction-endonuclease-digested DNA is annealed to segments of chromosomes that are fixed to a solid support and then washed to remove non-successfully hybridized fragments; the hybridized fragments are eluted and PCR-amplified. (*see also* coincidence painting)

• Hozier, J., Graham, R., Westfall, T., et al. (1994) Genomics **19**, 441–447

preparative ultracentrifugation The separation of subcellular organelles by high-speed centrifugation.

prep-ISH (= preparative *in situ* hybridization)

preproprotein The entire polypeptide that is encoded by an mRNA for a secretory protein before processing by proteinases cleaves the signal sequence and another polypeptide sequences that must be removed before the protein becomes functional. The sequence is preproprotein → proprotein (inactive) → protein (active). (*see also* prohormone; zymogen)

pre-synaptic Descriptive of a nerve cell that transmits its signal to another nerve cell across a synapse. (*see also* post-synaptic)

Pribnow box (= TATA box)

primary structure The amino acid sequence of a protein or polypeptide, or the nucleotide sequence of a polynucleotide. (*see also* quaternary structure; secondary structure; supersecondary structure; tertiary structure)

primer An RNA sequence hybridized to a DNA template whose elongation by a DNA polymerase constitutes DNA synthesis. A *random primer* is a mixture of polynucleotides with all four bases at each sequence position; an *arbitrary primer* is a single species with a single base at each sequence position.

primer extension analysis A method for measurement of the amount of a specific mRNA and its length relative to a known sequence. A labelled synthetic primer designed for complementarity to a sequence in the mRNA, usually close to the 3′-end, is reverse transcribed to generate a cDNA that by the intensity of its labelling indicates the amount of mRNA originally present, and by its mobility on gel electrophoresis indicates its length, and hence the length of the mRNA from the site of primer attachment to its 5′-end.

primer mismatch analysis A method for detection of the presence of a specific DNA fragment. Amplification by PCR with, alternatively, a wild-type-specific or a mutant-specific primer for the same chain and an anti-sense chain primer generates an amplification product only where there is a perfect match of primers to target.

• Dockhorn-Dwornaiczak, B., Schroder, S., Caspari, S., et al. (1991) in PCR Topics: Usage of Polymerase Chain Reaction in Genetic and Infectious Diseases (Rolfs, A., Schumacher, H.C. and Marx, P., eds.), pp. 224–229, Springer-Verlag, Berlin

primer walking A method for sequencing relatively long pieces of DNA by use of a library of short, and therefore inexpensive, polynucleotides to assemble a primer complementary to a known sequence, initially a vector sequence. Three or four appropriate hexanucleotides are assembled on a 18- or 24-base sequence of the template with suppression of the binding of individual hexanucleotides by competition from a single-stranded DNA-binding protein. The downstream sequence is determined as far as is practical, and a newly identified sequence is used to select and bind three or four new hexanucleotide primers for identification of the subsequent downstream sequence.

- Kieleczawa, J., Dun, J.J. and Studier, F.W. (1992) Science **258**, 1787–1791

primosome A complex of proteins that first synthesizes the RNA primer for DNA replication that is subsequently elongated to form an Okazaki fragment. (*see also* transcriptional activation)

prion A proteinaceous infective agent that causes one of a number of degenerative neurological diseases, e.g. bovine spongiform encephalopathy, scrapie in sheep and Creutzfeld–Jakob disease in humans. Unlike other infectious agents it is apparently composed entirely of a single protein, PrP^{Sc}, that differs from a normal cellular protein, PrP^{c}, only in its conformation; the benign conformation is converted into the infectious conformation when the two come together.

- Beyreuther, K. and Masters, C.L. (1994) Nature (London) **370**, 419–420; Smith, C. and Collinge, J. (1995) Essays Biochem. **29**, 157–174; Baldwin, M.A., Cohen, F.E. and Prusiner, S.B. (1995) J. Biol. Chem. **270**, 19197–19200

proband (= propositus)

probe (= hybridization probe)

processing Post-transcriptional modification of RNA or post-translational modification of a polypeptide or protein.

processivity The property of a polymerase to act continuously without dissociation from the template.

prochirality The symmetry characteristic of a molecule, or of a centre within a molecule, by which like groups may be distinguished, e.g. the property of citric acid that allows aconitase to distinguish

between its two carboxymethyl groups. (*see also* chirality; CIP classification; *meso*-carbon; *R*; symmetry)

procollagen suicide The magnified effect of a mutation in the propeptide of one collagen precursor molecule on the structure of secreted collagen. The phenomenon results from the disruption of fibril assembly when a small proportion of the subunits can form abnormal cross-links, e.g. disulphide bridges.

- Hulmes, D.J.S. (1992) Essays Biochem. **27**, 49–67

proenzyme (= zymogen)

progenote In early evolution, an organism with a rudimentary mechanism for replication and translation. (*see also* protogenote)

- Bennen, S.A. and Ellington, A.D. (1990) Science **248**, 943–944

progestogen A compound, usually a steroid, that prepares the uterus for implantation and supports pregnancy, e.g. progesterone.

programmed cell death (= apoptosis)

prohormone An inactive precursor of a hormone, especially a polypeptide hormone.

prokaryote (*see* eukaryote)

promoter A sequence of double-stranded DNA proximal to the 5'-end of the structural gene that regulates the binding and activity of RNA polymerase. (*see also* enhancer)

proneurophysin (*see* neurophysin)

proofreading (editing) In DNA replication, the 3'-to-5' exonuclease activity of a polymerase that removes incorrectly paired bases; in protein synthesis, the hydrolysis of an incorrectly selected amino acyl-tRNA by the amino acyl-tRNA synthetase. (*see also* excision)

proofwinding Analogous to proofreading, a putative activity that tests the accuracy of a transcription by hybridization of the product RNA to an authentic complementary polynucleotide.

propensity The likelihood of finding an amino acid residue in a certain secondary structure, e.g. glycine in an α-helix. Data from statistical surveys, host–guest analysis, model peptides, directed mutagenesis or molecular dynamics calculations are expressed as $\Delta\Delta G$ and quantified in kJ/mol (kcal/mol).

• Blaber, M., Zhang, X.-j. and Matthews, B.W. (1993) Science **260**, 1637–1640

propeptide The part of a protein that is proteolytically cleaved during the protein's maturation, e.g. the internal residues of proinsulin, the N-terminal residues of pepsinogen. The term is usually not used for a signal peptides, but for the peptides removed from protein precursors that can be isolated and that have a reasonably long half-life.

propinquity effect (*see* entropy effect)

propionate rule Analogous to the acetate and isoprene rules, the observation that some natural products are assembled from multiple propionate (propionyl-CoA and methylmalonyl-CoA) units in head-to-tail condensations. (*see also* macrolide; polyketide)

propositus (proband) In human genetics, the individual whose forebears and descendants are under investigation.

proprotein A precursor protein that requires post-translational modification for its activity to be expressed, e.g.

pepsinogen, trypsinogen. (*see also* pre-proprotein)

prosome (= proteasome)

prostaglandin An eicosanoid that features a cyclopentane nucleus.

prosthetic group A moiety that is tightly attached to a protein, often a participant in an enzymic reaction; e.g. haem, FAD, Zn^{2+}, pyridoxal phosphate. (*see also* apoprotein; holoprotein)

protamine A small, helical, arginine-rich basic protein that lies along the major groove of B-DNA.

protease (= endopeptidase)

proteasome A multicatalytic proteinase; a 700 000 Da tightly bound complex of several proteins and possibly polyribonucleotides that possesses several discrete latent and active endoproteinase activities.

• Skilton, H.E., Eperon, I.C. and Rivett, A.J. (1991) FEBS Lett. **279**, 351–355; Rechsteiner, M., Hoffman, L. and Dubiel, W. (1993) J. Biol. Chem. **268**, 6065–6068; Peters, J.-M. (1994) Trends Biochem. Sci. **19**, 377–382

protection footprinting (*see* footprinting)

protein A linear polymer of α-amino acids held together by peptide bonds. A protein may include other components attached covalently (e.g. carbohydrate, phosphate, fatty acid, biotin) or non-covalently (e.g. Zn^{2+}, FAD, haem), and may be cross-linked by disulphide bridges or, less commonly, by other specialized bonds. It may exist as a complex of two or more identical or dissimilar polypeptide chains.

proteinase (= endopeptidase)

protein cage The reaction site within which inorganic materials are deposited and that can reach the supramolecular (i.e. nanometer) size, e.g. apoferritin, within which hydrated ferric oxide is deposited to form ferritin.

• Meldrum, F.C., Wade, V.J., Nimmo, D.L., et al. (1991) Nature (London) **349**, 684–687

protein kinase An enzyme that uses ATP to phosphorylate a group on a protein, e.g. a serine, threonine or tyrosine hydroxy group. (*see also* MAP kinase; protein kinase C; tyrosine kinase)

protein kinase C (PKC) A calcium-dependent protein kinase; often one stimulated by diacylglycerol.

protein kinesis The directed cellular movement of proteins, e.g. nuclear proteins to the nucleus, lysosomal proteins to the lysosomes, etc.

• Hurtley, S.M. (1996) Science **271**, 1477; Rothman, J.E. and Wieland, F.T. (1996) Science **272**, 227–234

protein ladder sequencing Determination of a protein's amino acid sequence by the generation of a series of fragments from the original protein which represent sequences of single amino acid removals. The series is generated by stepwise degradation effected by Edman degradation using phenylisothiocyanate and a small amount of a termination reagent, phenylisocyanate. Cycles of reaction are performed without isolation of intermediates, and the products are characterized by mass spectrometry.

• Chait, B.T., Wang, R., Beavis, R.C. and Kent, S.H.B. (1993) Science **262**, 89–92

protein processing Limited proteolysis that converts a biologically inactive translation product into an active enzyme, hormone, structural protein, etc.

protein quality The relative dietary value of a protein that reflects its content of the various essential amino acids.

protein ruler A high-molecular-mass protein that functions in organization of cytoskeletal structures of uniform length by serving as a template for aggregation of exact numbers of other components of the structure along its length, e.g. nebulin, which is suggested to align the subunits of skeletal muscle thin filaments.

• Labeit, S., Gibson, T., Lakey, A., et al. (1991) FEBS Lett. **282**, 313–316

protein splicing The excision of an intervening protein sequence, an *intein*, from a precursor protein, with peptide bond formation between the flanking domains (the *N-extein*, originally attached to the N-terminal end of the intein, and the *C-extein*, originally attached to its C-terminal end) to produce the spliced protein. Intein sequences are homologous with homing endonucleases that transfer the intron gene to an intron-less allele, and some free intein proteins possess the endonuclease activity. The intein–C-extein interface is an asparagine–serine bond. The splicing mechanism is autocatalytic, and is proposed to employ displacement of the N-extein from the intein by the β-hydroxy group of the serine. This produces a branched intermediate: to the C-extein serine residue are attached the intein in a peptide bond to the α-amino group and the N-extein in an ester bond to the β-hydroxy group.

The asparagine residue cyclizes to free the intein, with a C-terminal aminosuccinimide residue formed from the asparagine; the N- and C-exteins remain joined by an ester bond. Displacement of the N-extein from the β-hydroxy group by serine's α-amino group produces the spliced protein. Some C-exteins have threonine or cysteine in place of the serine.

• Cooper, A.A. and Stevens, T.H. (1995) Trends Biochem. Sci. **20**, 351–356

proteodermochondran sulphate A polymer of the extracellular matrix, derived from chondroitin 4-sulphate that is incompletely epimerized and sulphated. The completely processed polymer is dermatan sulphate. (*see also* glycosaminoglycan; proteoglycan)

• Scott, J.E. (1994) The Biochemist (Bull. Biochem. Soc.) **16(2)**, 32–34

proteoglycan (mucopolysaccharide) A complex composed of glycosaminoglycans that radiate from a protein core in a bottlebrush-like structure. There are two major types of membrane proteoglycans: *glypicans*, which are anchored to the peripheral surface by a glycosylphosphatidylinositol linkage; and *syndecans*, which contain transmembrane regions and short cytoplasmic tails. (*see also* core protein; link protein)

proteolipid A protein that is soluble in organic solvents, usually due to the presence of a significant number of fatty acids esterified to secondary hydroxy groups and/or fatty acylation of its N-terminus.

proteome Analogously to the genome, the protein composition of a cell.

• Kahn, P.J. (1995) Science **270**, 369–370

protist A unicellular eukaryote.

proto- Borrowing from historical linguistics, a prefix that denotes the simplest possible precursor of two or more organisms, inferred from phylogenetic trees reconstructed from polynucleotide sequences, e.g. protogenote (but not proto-oncogene).

• Bennen, S.A. and Ellington, A.D. (1990) Science **248**, 943–944

protofilament The linear subunit of a microtubule; composed of alternating α- and β-tubulin subunits. Thirteen protofilaments arranged into a cylinder constitute a microtubule.

protogenote In early evolution, the most recent common ancestor of cellular organisms (which include the archaebacteria, eubacteria and eukaryotes) with only rudimentary translation capabilities. (*see also* progenote)

• Popper, K.R. and Wachtershauser, G. (1990) Science **250**, 1070

protome The basic subunit of a multimeric protein.

proton inventory (*see* solvent isotope effect)

proton motive force The free energy, expressed in volts, represented by a pH gradient across the membrane of a vesicle due to the concentration gradient plus the membrane potential. This energy is used by coupling the transport of electrons across the membrane with the synthesis of ATP. (*see also* chemiosmotic theory)

proton motive Q cycle The cyclic reduction and oxidation of cytochromes b and c_1 and an iron–sulphur protein, which comprise the cytochrome bc_1 complex. In each cycle one ubiquinol is oxidized to ubiquinone, two cytochrome c_{ox} are reduced to cytochrome c_{red} and four protons (two oxidized from ubiquinol and two from water) are transported from the inner, electronegative, surface to the outer, electropositive, surface of mitochondria and of many bacteria.

• Trumpower, B.L. (1990) J. Biol. Chem. **265**, 11409–11412

protonophore (*see* ionophore)

proton pump A mechanism for the active transport of a proton across a membrane, e.g. the proton pump of the mitochondrial inner membrane, of the gastric parietal cell, of the thylakoid membrane.

proton switch In enzyme chemistry, a mechanism of tautomerization in which a proton is removed by a solvent molecule from an ionizable intermediate in the catalytic cycle and added to another atom of the intermediate from another solvent molecule.

proto-oncogene A normal cellular gene that has the potential to become an oncogene by mutation or translocation; identical with c-oncogene. (*see also* anti-oncogene)

protoplast A plant cell that has been stripped of its cell wall so as to be limited only by its plasma membrane. (*see also* spheroplast)

provirus A stage in the development of a virus, in which it is incorporated into the genome of a host cell; in a retrovirus, the double-stranded DNA stage.

proximity effect (*see* entropy effect)

pseudo-first order In kinetics, descriptive of a rate that is directly dependent upon the concentration of one reactant that is experimentally varied. The rate may also be dependent upon the fixed concentrations of other reactants.

pseudogene A DNA sequence that is homologous to a structural gene, but cannot be expressed because it has no continuous open reading frame. It often occurs without introns.

pseudoglobulin (*see also* globulin)

pseudo-intron A sequence of an RNA transcript that sometimes behaves like an intron and is spliced out, but sometimes remains in the mature mRNA. (*see also* alternative splicing; read-through)

pseudo-knot RNA (*see* ribozyme)

pseudomutant The product of a silent mutation.

pseudopod A rounded projection of an animal cell used for locomotion. The cytoplasm streams into the pseudopod and extends the cell in its direction.

psi (Ψ) A symbol that in a designation of an amino acid sequence denotes a structure that substitutes for a peptide bond, e.g. Gly-Ψ(CH_2-CO)-Gly is H_2N-CH_2-CH_2-CO-CH_2-COOH.

psi-angle (Ψ-angle) A characteristic of a peptide bond; the angle formed at the α-carbon by the bonds to the nitrogen atom and the carbonyl carbon atom. (see *phi*-angle; Ramachandran plot)

PS-SPCL Positional scanning synthetic peptide combinatorial library. (*see* synthetic peptide combinatorial library)

psychrophile (*see* extremophile)

puckering The divergence from planarity of a non-aromatic ring compound, e.g. a sugar or steroid, to accommodate the bond angles of the constituent atoms. (*see also* chair form; half-chair conformation)

puff A local expansion of a chromosome

due to the unfolding of DNA as it is being transcribed.

pulse–chase experiment (*see* pulse labelling)

pulsed amperimetric detection (PAD) A method for detection of chromatographic eluates, especially oligosaccharides, not easily detected by other means; a cycle of less than 1 s first applies a voltage across the eluate, adsorbs the carbohydrate to an electrode and oxidizes it (the oxidizing current is what is monitored), followed by a strong oxidizing voltage which cleans the electrode for the start of the next cycle.

pulsed-field gel electrophoresis (PFGE) A technique for separation of chromosomes and very large DNA molecules, usually for purposes of genetic mapping; alternating pulses of electricity directed at 120° across the plane of an agarose gel slab drive the structures through the gel.

• Chu, G. (1990) Methods Companion Methods Enzymol. **1**, 129–142

pulse labelling In metabolic studies, the exposure of a biological system (e.g. a tissue slice, a cell culture, mitochondria) to a radiolabelled metabolite over a relatively short time, in order to subsequently follow the passage of the label from precursor metabolite to product metabolite. A variant is *pulse–chase labelling*, in which the pulse is terminated by the addition of a large amount of the unlabelled metabolite in order to rapidly lower its specific radioactivity. (*see also* equilibrium labelling)

pump An energy-dependent mechanism by which a metabolite or ion is forced across a membrane against a concentration gradient.

punctuation In translation, the polynucleotide sequence that signals the initiation or termination of a message.

purine One of the two classes of heterocyclic organic bases that are found in nucleic acids (principally adenine and guanine) and in several other kinds of biological compounds, e.g. coenzymes, nucleotides, sugar derivatives; also includes methylated purine alkaloids such as caffeine and theobromine. (*see also* pyrimidine)

purple membrane protein A constituent of *Halobacterium halobium*; a retinal-containing protein that traps light and transports protons across the membrane in which it is embedded.

putrefaction The process of decay of organic material by bacterial, fungal or non-biological processes into foul-smelling products.

pyranose The form of a sugar when it is condensed into a six-membered ring, which consists of five carbon atoms and the oxygen atom that is the link to the anomeric carbon. (*see also* furanose)

pyridine nucleotide NAD^+, NADH, $NADP^+$ or NADPH.

pyrimidine One of the two classes of heterocyclic organic bases that are found in nucleic acids (principally uracil and cytosine in RNA; uracil and thymine in DNA) and in several other kinds of biological compounds, e.g. nucleotides, sugar and lipid derivatives. (*see also* purine)

pyrimidine dimer A covalent attachment of two adjacent bases that can form when ultraviolet light damages double-stranded DNA; e.g. thymidine dimer.

Q

q In genetics, a designation of the short arm of a chromosome. (*see also* centromere)

Q_{10} The factor by which a reaction is accelerated upon raising the temperature by 10 degrees. (*see also* activation energy)

Qβ replicase A viral RNA-dependent RNA polymerase; used to greatly amplify concentrations of an RNA that serves as a template.

Q_{CO_2} The rate of production of CO_2 by a preparation, expressed as $\mu l/h$ per tissue dry weight. (*see also* manometry)

QC-PCR Quantitative competitive PCR. (= quantitative PCR)

Q_{O_2} The rate of uptake of O_2 by a preparation, expressed as $\mu l/h$ per tissue dry weight. (*see also* manometry)

QPCR (= quantitative PCR)

quad (*see* enzyme mechanism)

quantitative competitive PCR (= quantitative PCR)

quantitative PCR (QPCR) A method for quantification of a polynucleotide by inclusion with it of a known amount of an easily distinguished template as an internal standard to compensate for variation in efficiency of amplification. At the termination of cycling, the relative amounts of the products are measured by ethidium bromide binding and fluorescence or by incorporation of ^{32}P-labelled primers.

• Foley, K.P., Leonard, M. and Engel, J.D. (1993) Trends Genet. **9**, 380–385

quantum yield The ratio of the number of molecules that respond, e.g. by reacting or fluorescing, to the number of photons absorbed.

quaternary structure The arrangement in space of polypeptide subunits that make up a multimeric protein. (*see also* primary structure; secondary structure; supersecondary structure; tertiary structure)

quenching In fluorescence spectroscopy, a decrease in emission due to a variety of causes that dissipate energy of the excited state, e.g. transfer of energy to a non-fluorescent molecule such as molecular oxygen.

quiescent affinity label A reagent that reacts specifically with the active site of the enzyme for which it was designed. It bears a generally reactive chemical group, such as a chloromethane, and substrate-like moieties that permit it to bind only to the target enzyme. (*see also* affinity labelling)

quinimine form A tautomer of the Schiff base adduct of pyridoxal phosphate with an amino acid in which the pyridine ring nitrogen is protonated but uncharged and the carbon atom of the coenzyme that attaches the amino group of the amino acid has lost a proton. It is an intermediate between the *aldimine form* (in which the pyridine ring is protonated and charged and the attachment of the amino acid to the cofactor is a Schiff base) and the *ketimine form* (which can best be represented as a Schiff base formed by condensation of an α-oxo acid with pyridoxamine).

quinoprotein A protein, usually of microbial origin, with a quinone prosthetic group: pyrroloquinoline quinone (e.g. methanol dehydrogenase), 6-hydrodopaquinone (topa quinone, e.g. amine oxidase) or 2′,4-bitryptophane-6,7-dione (tryptophan tryptophanylquinone, e.g. amine dehydrogenase).

• McIntire, W.S. (1994) FASEB J. **8**, 513–521; Klinman, J.P. and Myu, D. (1994) Annu. Rev. Biochem. **63**, 299–344

quinozyme (= quinoprotein)

R

R A designation of absolute configuration at an asymmetrical centre; classified as *R* (*rectus*) or *S* (*sinister*) according to the arrangement in space of its substituents, which are assigned precedence according to sequence rules. Highest priority is given to the highest atomic number. If for an asymmetrical carbon atom, when viewed from the direction opposite the substituent of lowest priority, the sequence from the highest to lowest priority of the remaining substituents follows a clockwise path, the configuration is *R*; if counter-clockwise it is *S*. (*see also* chirality; CIP classification; prochirality)

RACE (= rapid amplification of cDNA ends)

RACE cleavage (= RecA-assisted cleavage by endonuclease)

racemic mixture The mixture in equal amounts of two enantiomers, e.g. D- and L-alanine.

rack model (= induced-fit theory)

radiation-induced hybrid mapping A method for location of a human genetic marker within a restricted region of a single chromosome. A hybrid cell line that contains a specific single human chromosome in each recipient rodent cell is heavily irradiated, and surviving cells are selected that carry fragments of the human chromosome integrated into the rodent chromosome. Each rodent chromosome is subsequently characterized by its own morphology and genetic markers and by the human fragment it carries, so that the cells may be analysed,

e.g. by fluorescence *in situ* hybridization, for a new genetic marker.

• Walter, M.A. and Goodfellow, P.N. (1993) Trends Genet. **9**, 352–356

radiationless transfer The fall to the ground state of an excited molecule without the emission of light, i.e. without fluorescence, so that the energy appears as heat.

radioautography (= autoradiography)

radioimmunoassay (RIA) A competitive binding assay that depends upon displacement of a radiolabelled ligand from an antibody by the non-labelled standard or test sample, followed by separation of the labelled ligand into bound and unbound fractions. (*see also* ELISA)

Ramachandran plot A graph that displays calculated energy isotherms for conformations of peptide backbones; *phi*-angles against *psi*-angles; energy minima correspond to β-structure, α-helix and the collagen triple helix.

Raman optical activity (*see* vibrational optical activity)

Raman scattering (*see* light scattering)

random (*see* enzyme mechanism)

random coil The unfolded, denatured, state of a protein or nucleic acid.

randomization The migration of an isotope from one position in a labelled metabolite to another position due to the conversion of the metabolite into a symmetrical compound, in which labelled and non-labelled positions become equivalent, or by a process in which moieties of a metabolite are transformed into one another; e.g. the fate of the carboxy and methyl carbons of acetate as it passes through the reactions of the tricarboxylic acid cycle.

random oligonucleotide mutagenesis (= saturation mutagenesis)

random peptide library A collection of bacteria or viruses, engineered to bear randomly mutated coat proteins. The bacteria or viruses are screened for high-affinity binding to a target ligand; the bacterium or virus that bears the desired binding characteristics and the DNA sequence that codes for them are thereby selected for growth and characterization.

• Scott, J.K. and Smith, G.P. (1990) Science **249**, 386–390; Scott, J.K. (1992) Trends Biochem. Sci. **17**, 241–245

random primer (*see* primer)

random walk The irregular path followed by a particle that shows Brownian movement or by a flagellar bacterium that periodically changes its direction of motion.

rapid amplification of cDNA ends (RACE) A PCR-based method to obtain cDNA clones of rare transcripts which begin with only a partial internal sequence. The 3′- and 5′-ends are analysed separately; a single (−)-strand is first constructed from the 3′-end of the transcript by use of reverse transcriptase and a poly(dT) primer. The (+)-3′-end is then generated in multiple copies by PCR using poly(dA) and the known internal partial sequence as primers. A single (−)-strand copy is constructed from the 5′-end using reverse transcriptase with a (−)-primer based on the known internal sequence. This 5′-(−)-transcript is tailed with poly(dA) and then used as a template in a PCR with the known internal sequence and poly(dA) as primers. Once the 3′- and 5′-ends are known, primers can be constructed to produce multiple copies of the gene by PCR. (*see also* RecA-assisted cleavage by endonuclease)

• Frohman, M.A. (1990) in PCR Protocols: Rapid Amplification of cDNA Ends (Innis, M.A., Gelfand, D.H., Sninsky, J.J. and White, T.J., eds.), pp. 28–38, Academic Press, San Diego

rapid mixing (*see* stopped-flow)

RAP-PCR RNA fingerprinting arbitrary primer PCR. (*see* differential PCR display)

rare-cutter A restriction nuclease that is specific for nucleotide sequences that occur infrequently in the genome.

ras **gene** An oncogene of the murine sarcoma virus. (*see also* ras protein)

ras protein A product of the *ras* gene; a G-protein.

rate-limiting step The slowest step of a metabolic pathway or enzymic reaction; the one that determines the rate of appearance of the ultimate product.

Rayleigh scattering (*see* light scattering)

R band (*see* AT queue)

RCS Recognition consensus sequence. (*see* shape library)

RDA (= recommended daily allowance, or representational difference analysis)

rDNA Recombinant DNA.

reaction centre In photosynthesis, the energy-transducing unit. By absorption of light it excites a chlorophyll dimer, which can then transfer a pair of electrons across a membrane to a quinone, which will accept two protons from its microenvironment to generate a hydroquinone.

reaction co-ordinate A graphical representation of the course of a reaction that shows the energy of the reactants and products as the reaction progresses.

reactive immunization A technique to widen the scope of catalytic reactions effected by abzymes. Reactive compounds are used as antigens; some of the resulting antibodies are directed to the reaction complex with a protein or carbohydrate. The antibody can then recruit to its binding site the components of the reaction and force them into reactive juxtaposition.

• Wirsching, P., Ashley, J.A., Lo, C.-H., et al. (1995) Science **270**, 1175–1182

reactive oxygen The group of chemically highly reactive species that are related to molecular oxygen, i.e. $O_2^{-\bullet}$ (superoxide), $HO^{\bullet}$ (hydroxyl radical) and H_2O_2 (hydrogen peroxide).

reading frame The register in which the translation apparatus senses the information coded within an mRNA molecule. As the code is a triplet, there are three possible reading frames. (*see also* open reading frame)

read-through A protein that is the product of alternative splicing of the transcription product. A potential intron remains in the mRNA and is translated rather than excised. (*see also* pseudo-intron)

RecA-assisted restriction cleavage by endonuclease (RACE cleavage) A method for cleavage of double-stranded DNA at specific restriction sites. A suitable restriction site is selected, around which the base sequence is known, and an oligonucleotide, up to 50 or 60 bases in length, is synthesized to complement it. It is hybridized to the DNA to form a triple helix, which is stabilized with the help of the RecA protein. Methylation then blocks all but the protected restriction site, so that subsequent treatment with the restriction nuclease, in the absence of the protecting oligonucleotide and RecA protein, will cleave only at the designated site. The use of two protecting oligonucleotides for sites that span a sequence of interest allows excision of a specific fragment from a DNA preparation as complex as the human genome.

• Ferrin, L.J. and Camerini-Otero, R.D. (1991) Science **254**, 1494–1497

receptor The binding site for a hormone or neurotransmitter that initiates its action at the cellular level.

receptosome (endosome) A coated vesicle from which the clathrin shell has been removed.

recessive (*see* allele)

recoding The altered reading of an mRNA due to instructions (recoding signals) embedded in the sequence, which cause frame shifts or other reading alterations.

• Gesteland, R.F., Weiss, R.B. and Atkins, J.F. (1992) Science **257**, 1640–1641

recognition consensus sequence (*see* shape library)

recombinant DNA DNA formed by bringing together DNA fragments from different species.

recombinant PCR A protocol for site-directed mutation and construction of a vector containing the mutation. A pair of homologous complementary primers that contain the mutation are used separately with other primers each of which, with its mutagenic primer, bracket a unique restriction site in order to construct two double-stranded oligonucleotides that overlap only over the sequence of the mutagenic primers. These double-stranded DNAs are mixed, denatured and annealed to form some of a hybrid that is double-stranded only in the sequence of the mutagenic primers and single-stranded in the direction that extends towards the other external primer sites. PCR amplification and restriction fills in the full-length DNA and prepares it for insertion into a vector.

• Higuchi, R. (1990) in PCR Protocols: Rapid Amplification of cDNA Ends (Innis, M.A., Gelfand, D.H., Sninsky, J.J. and White, T.J., eds.), pp. 177–183, Academic Press, San Diego

recombination The natural or synthetic production of a DNA molecule from polynucleotides derived from more than one parent DNA molecule.

recombination joint The site of interaction between two double-stranded DNA molecules; at regions where the two double strands are homologous, equivalent strands of each duplex break and exchange complementary strands to form a bridge between the two DNAs. (*see also* Holliday model)

recommended daily allowance (RDA) The intake of nutrients, vitamins, minerals and protein considered advisable by the Ministry of Agriculture, Fisheries and Food (U.K.), or the Food and Nutrition Board of the National Academy of Sciences or the Food and Drug Administration (U.S.A.); usually greater than the minimum daily requirement.

red drop The sharp decline in photosynthetic efficiency as the wavelength of exciting light enters the red region (above 680 nm) due to the insensitivity of photosystem II to excitation by red light.

red muscle (slow-twitch muscle) A well vascularized form of skeletal muscle with many mitochondria and much cellular myoglobin; supplied with energy mainly by β-oxidation. (*see also* white muscle)

redox potential ($E^{0'}$) The electrical potential, expressed in volts, of a reductive half-reaction, e.g. for $NAD^+ + 2e^- + H^+ = NADH$, $E^{0'} = -0.32$ V.

reducing sugar A sugar whose anomeric carbon atom can be oxidized by an alkaline cupric ion solution, e.g. Fehling's solution, Benedict's solution. (*see also* non-reducing sugar)

reduction In its essential form, the addition of electrons to a compound; possibly with the addition or abstraction of other atoms.

reduction division The first phase of meiosis, in which a diploid cell gives rise to two haploid daughter cells.

reduction potential (= redox potential)

reductive pentose cycle (= Calvin cycle)

REF (= restriction endonuclease fingerprinting)

reflectional symmetry (*see* symmetry)

refractive index The ratio of the velocity of light in a vacuum to the velocity in the medium in question.

regioselective (*see* regiospecific)

regiospecific Descriptive of a reaction that proceeds by a unique course even though more than one course is formally possible, e.g. an addition that occurs at only one atom of an unsymmetrical olefin; contrasted with *regioselective*, in which the reaction shows merely a preference for one course over another; and with *stereospecific*, in which the reaction produces one diastereoisomer rather than another.

regulability (*see* regulation)

regulation The phenomenon of adjustment of metabolite flux, e.g. glucose that enters the anaerobic glycolysis pathway and exits as lactate, whatever its molecular mechanism. It reflects for each step of the pathway the range over which that step's activity can be modulated (*regulability*) and the effect this has on overall metabolic flux (*control*). Flux-control, metabolite-control, elasticity and response coefficients have been used to model and analyse the effects of substrate, inhibitor and allosteric effector concentrations.

• Slater, M., Knowles, R.G. and Pogson, C.I. (1994) Essays Biochem. **28**, 1–12

regulatory subunit A part of a multimeric enzyme; a protein that binds to and inhibits the *catalytic subunit* unless it is bound to a regulating molecule; e.g. in the case of cyclic AMP-dependent protein kinase, the subunit that, unless bound to the cyclic nucleotide, inhibits the catalytic subunit.

reiteration Repetition of identical, or nearly identical, long DNA sequences in a chromosome.

relative contact number (*see* contact number)

relaxation In kinetics and spectroscopy, the return to equilibrium of a macromolecule after a very brief perturbation of the environment, e.g. recovery of a protein from a sudden burst of energy that allows a fast transient increase in temperature

(*temperature-jump*) or a fast transient change of pH (*pH-jump*).

relaxed Descriptive of the topology of closed circular DNA with no superhelical torsion applied to it, which has one turn per 10.4 base pairs.

relaxed bacteria (*see* magic spot nucleotides)

relaxed conformation R-form. (*see* Monod–Wyman–Changeux model)

release factor A protein that binds to the terminator codon of mRNA and allows hydrolysis of the peptidyl-tRNA ester bond.

releasing factor (= releasing hormone)

releasing hormone A hormone, often of hypothalamic origin, whose function is to cause release of another hormone, often of hypophyseal origin; e.g. thyrotropin-releasing hormone.

renaturation The return to native structure from a denatured state; in nucleic acid chemistry, identical to annealing.

renin An aspartate proteinase that is secreted by the kidneys and releases angiotensin I from angiotensinogen in the blood; not to be confused with rennin.

rennin Obsolete name for chymosin, a milk-clotting aspartate proteinase secreted by the stomachs of newborn mammals; not to be confused with renin.

repeat A feature of a helix; the distance parallel to the axis in which the backbone structure repeats itself. For a DNA double helix, for example, the repeat distance is 34 Å (0.34 nm), which represents 10 residues. (*see also* pitch; rise)

repertoire In molecular immunology, the assortment of immunoglobulins produced by an animal; this assortment changes (*repertoire shift*) as the animal develops immunity following a challenge.

repertoire cloning A method to produce human monoclonal antibodies. mRNAs are isolated from a tissue that generates immunoglobulins (e.g. spleen) and a reverse transcriptase makes cDNA copies of them. The Fd regions, i.e. the H-chain component of the Fab fragment (the V_H and C_H1 domains of the immunoglobulins regions), are selected and amplified by PCR to obtain a *heavy-chain library*. Similarly, a *light-chain library* is obtained. The two populations of cDNA are combined into a random pairing to form combinatorial constructs, inserted into coliphage vector along with a promoter, and used to infect *E. coli*. Plaques are then screened for ability to bind an antigen. (*see also* synthetic peptide combinatorial library)

• Barton, D.R. (1991) Trends Biotechnol. **9**, 169–175

repertoire shift (*see* repertoire)

repetitive DNA Sequences of DNA that are found to be repeated, sometimes thousands of times over.

replication Synthesis of new DNA strands complementary to existing template strands.

replication fork The site where the two polynucleotides of a parent DNA separate during replication. The daughter strands are attached to the two separated polynucleotides that trail away from the fork as it advances into the parent DNA. (*see also* Okazaki fragment)

replication licencing factor (*see* licencing)

replicon A segment of the eukaryotic genome that contains several genes and is replicated as a unit from a single origin. Replication is bi-directional; its boundaries are the points where replication from one origin meets the replication fork which advances from the opposite direction. Each replicon is presumed to be identical to a DNA loop.

reporter group A chromophore or fluorophore that is sensitive to its environment; a group added to a protein to indicate changes in its environment as the protein is manipulated, e.g. by denaturation or association with a ligand or other proteins.

representation A portion of the genome with lowered complexity (e.g. by cleavage with an infrequent cutting restriction nuclease, ligated to oligonucleotide adapters so it can then be amplified by PCR), but that can be used in subtractive cloning and kinetic selection procedures. (*see also* representational difference analysis)

• Lisitsyn, N.A. (1995) Trends Genet. **11**, 303–307

representational difference analysis (RDA) A method for isolation and amplification of DNA sequences that differ between two populations (a subtractive technique); e.g. changes associated with antigenic variation in micro-organisms, activation of the immune system, chromosomal translocation. The complexity of one population, the *tester* DNA, is lowered by cleavage with an infrequent cutting endonuclease, ligated to adaptor oligonucleotides to allow PCR amplification; the other population, the *driver*

DNA, is similarly treated, but after amplification the ligated adaptors are removed to prevent amplification. These *amplicons* do not include the entire parent population, but serve as representations. Tester and an excess of driver DNA are mixed, denatured and annealed. By using kinetic enrichment conditions, the tester strands common to the two populations will be annealed to driver strands; only sequences unique to the tester population will form duplexes. The ends of the duplexes are filled in with a polymerase and PCR amplified; only fragments found uniquely in the tester population will amplify exponentially, as hybrid duplexes undergo only linear amplification and therefore will be a small fraction of the amplified fragments.

• Lisitsyn, N.A. (1995) Trends Genet. **11**, 303–307

repressor A protein that can regulate transcription by binding to the operator and causing repression.

RER (= rough endoplasmic reticulum)

rescue A strategy in chemotherapy in which lethal doses of an antifolate are given to block nucleotide biosynthesis in rapidly proliferating cells, followed by treatment with 5-formyltetrahydrofolate or thymidine to 'rescue' normal cells.

residualizing label A moiety attached to a circulating protein in order to subsequently detect by histological means the site of the protein's degradation in a multicellular organism. A reporter group, such as a radioactive or fluorescent group that does not alter the protein's turnover rate, is attached to the protein under non-denaturing conditions, and is retained in lysosomes of those cells that degrade the protein.

residue The part of a single sugar that appears in a polysaccharide; of a single amino acid in a protein; of a single nucleotide in a nucleic acid, etc.; usually the monomer minus the elements of water.

resistance transfer factor (R-factor) A plasmid that contains genes for resistance to several antibiotics, which permits the transfer of drug resistance between bacteria.

resolution In X-ray crystallography, the precision with which atoms are located in space; usually expressed in Å (10^{-10} m).

resonance energy transfer A phenomenon in which one fluorophore excites another when there is significant overlap of the emission spectrum of the primary fluorophore with the excitation spectrum of the secondary fluorophore and some other conditions are met. In part it depends strongly upon the distance that separates the fluorophores and their orientation towards one another; it is useful in calculation of intermolecular distances of macromolecules. (*see also* molecular ruler)

• Clegg, R.M. (1995) Curr. Opin. Biotechnol. **6**, 103–110

resonance scattering (= anomalous scattering)

resorption Reabsorption; the reversal of an excretion, e.g. re-uptake of Na^+ in kidney tubules, or of a deposition, e.g. degradation of bone mineral by osteoclasts.

respiration The consumption of oxygen by an organism, organ, tissue, cell or subcellular structure, as a terminal electron acceptor in the conversion of metabolites into carbon dioxide.

respiration state The condition of tightly coupled mitochondria that is characterized by their respiratory rate (all but state 3 are slow) and defined by the rate-limiting component: state 1, ADP- and substrate-limiting; state 2, substrate-limiting; state 3, respiratory chain-limiting; state 4, ADP-limiting; state 5, oxygen-limiting.

respiratory burst The sudden consumption of oxygen by a leucocyte when it is stimulated.

respiratory chain (= electron transport chain)

respiratory control index A quantitative measure of the degree of coupling of phosphorylation to oxidation; the ratio of the rates of respiratory state 3 to state 4.

respiratory quotient (RQ) The molecular ratio of carbon dioxide generated to oxygen consumed.

resting metabolic rate (RMR) The rate of oxygen consumption by a person at rest which represents the energy required for essential functions such as respiration, circulation, maintenance of temperature. Also known as the basal metabolic rate (BMR).

resting potential The membrane potential of an excitable cell in the absence of stimulation.

restriction endonuclease One of a group of enzymes that cleave internal phosphodiester bonds of both strands of DNA at specific nucleotide sequences, especially at palindromic sequences. (*see also* isoschizomers)

restriction endonculease fingerprinting (REF) A method for analysis of large restriction fragments of DNA with a high degree of confidence of finding mutations. The large fragment is separately digested with several cocktails (preferably five) of endonucleases. The resulting digests are denatured, mixed together, labelled and electrophoresed on non-denaturing gels. In the restriction component of the analysis, additional or absent electrophoretic bands, compared with normal DNA, will result when a mutation creates, destroys or changes the electrophoretic mobility of a band.

• Liu, Q. and Somers, S.S. (1995) Biotechnology **18**, 470–477

restriction fragment A fragment of double-stranded DNA that results from the cleavage of DNA by a restriction endonuclease.

restriction fragment length polymorphism (RFLP) A phenomenon detected by Southern blotting; the variation within the population of the size of restriction fragments in which appear various genetic markers; e.g. any of the thalassaemias.

restriction map The characterization of double-stranded DNA by the locations of the sites at which various restriction endonucleases can cleave it. (*see also* contig map; linkage map; physical map; sequence-tagged site)

restriction-site-generating PCR (RG-PCR) A method for detection of a specific mutation that does not itself cause a unique restriction fragment length polymorphism, because it neither creates nor removes a known restriction site. A new restriction site is created by the introduction of a deliberately mismatched PCR primer, the 5'-end of which is designed to hybridize to the mutant template and the 3'-end of which encodes a new restriction site. (*see also* mismatched PCR)

• Ehrlich, G., Ginzberg, D., Loewenstein, Y., et al. (1994) Genomics **22**, 288–295

restrictive temperature For a heat-sensitive cell line, a temperature at which it cannot function.

retentate Non-filtered particles and macromolecules.

retention of configuration The lack of a change in the configuration at an asymmetrical centre over the course of a reaction, e.g. the enzymic addition of an α-glucosyl unit to glycogen from UDP-glucose, which is itself an α-glucoside. Such reactions may involve two Walden inversions.

reticulocyte A non-nucleated (in mammals) precursor of an erythrocyte, still capable of haemoglobin synthesis

reticuloendothelial system The tissues involved in synthesis and degradation of blood cells and their constituents; includes Kupffer, bone marrow and spleen cells.

retinoid A sesquiterpene related to or derived from retinol.

retroelement A mobile genetic element; either a retrovirus or a transposon, both of which are characterized by long terminal repeats.

• Temin, H.M. (1989) Nature (London) **339**, 254–255

retro-inverso Descriptive of a polypeptide analogue in which the direction has been inverted internally, giving it two N- or two C-termini, e.g. ...CO-CHRNH-CO-NHCHR'CO...

retrotransposon A mobile genetic element composed of repetitive DNA sequences that flank open reading frames, including one that codes for a reverse transcriptase; transposed by reverse transcription of the RNA into cDNA, its conversion into double-stranded DNA and insertion into a new position in the genome. (*see also* retroelement)

retrovirus (*see* RNA virus)

reverse genetics (= positional cloning)

reverse transcriptase A DNA polymerase that uses an RNA template; an RNA-dependent DNA polymerase.

reverse transcriptase-PCR (RT-PCR) Also known as RNA phenotyping, RNA-PCR or message amplification phenotyping; a method for amplification of a specific mRNA by prior use of reverse transcriptase to form a cDNA, then use of PCR to amplify. (*see also* competitive PCR)

• Larrick, J.W. (1992) Trends Biotechnol. **10**, 146–152

reverse turn A secondary structure of proteins in which the polypeptide backbone turns sharply on itself in as few as three residues, and is stabilized by hydrogen bonds between the amide hydrogens and carbonyl oxygens on either side of the turn. Also known as a β-bend, a β-turn or a hairpin bend.

R_F In thin layer and paper chromatography, the ratio of the distance moved by a test sample to the distance moved by the solvent front or, in gel electrophoresis, to the distance moved by a low-molecular-mass marker.

R-factor (= resistance transfer factor)

RFLP (= restriction fragment length polymorphism)

R-form Relaxed conformation. (*see* Monod–Wyman–Changeux model)

RG-PCR (= restriction-site-generating PCR)

rho protein (*see* anti-terminator; termination factor)

RIA (= radioimmunoassay)

ribophorin A protein of the rough endoplasmic reticulum that recognizes leader sequences of growing polypeptides and, through them, attaches the ribosomes.

riboside Ribonucleoside. (*see* nucleoside)

ribosomal RNA (= rRNA)

ribosomal stuttering A phenomenon in which translation, e.g. of a retroviral polyprotein message, compensates for a frame shift in the message itself.

ribosome A non-membrane-bound organelle; a complex of RNA molecules and proteins that is a site of protein synthesis in eukaryotes. (*see also* heavy subunit; light subunit)

ribosome-inactivating protein (RIP) One of a class of plant proteins that bind to elongation factor 2 and thus inactivate ribosomes. Type 1 (e.g. the toxin gelonin) consist of single polypeptide chain proteins; type 2 (e.g. ricin) consist of two proteins linked by a disulphide bond, one the toxin and the other a lectin that attaches to recognition sites on a target cell.

ribotide Ribonucleotide. (*see* nucleotide)

ribozyme An RNA molecule that, due to peculiarities in its secondary structure, is able to catalyse the interchange of some of its own phosphodiester linkages to achieve intramolecular splicing. Examples are *hammerhead*, *hairpin* and *pseudo-knot* RNAs, whimsically named for their topologies, that normally cleave

cis, i.e. intramolecularly, but can be engineered to cleave *trans*, i.e. intermolecularly, a designated RNA target. (*see also* intein; self-splicing)

• Been, M.D. (1994) Trends Biochem. Sci. **19**, 251–256; Cech, T.R. and Uhlenbeck, O.C. (1994) Nature (London) **372**, 39–40

right-handed helix By analogy with the right hand, a helix, e.g. of a polypeptide, that advances towards the C-terminus in the direction of the extended thumb as the backbone of a chain turns in the direction of the fingers closing on the palm. A *left-handed helix* has the opposite sense.

RIP (= ribosome-inactivating protein)

rise A feature of a helix. The distance parallel to the axis corresponds to one residue, e.g. 1.5 Å (0.15 nm) for an α-helix. (*see also* pitch; repeat)

RLF Replication licencing factor. (*see* licencing)

R loop (*see* AT queue)

RL-RT-PCR (= RNA ligase reverse transcription PCR)

RMR (= resting metabolic rate)

RNA Ribonucleic acid; a macromolecule formed of repeating ribonucleotide units linked by phosphodiester bonds between the 5'-phosphate group of one nucleotide and the 3'-hydroxy group of the next. RNA has several biological functions, most of which depend upon its ability to form sequence-specific interactions with DNA. RNA comprises the genome of some viruses. (*see also* mRNA; rRNA; tRNA)

RNA chaperone A putative RNA-binding protein that assists RNA in attainment of its biologically active conformation.

Non-specific RNA-binding proteins are proposed to prevent RNA becoming temporarily trapped in innumerable non-functional metastable conformations (*kinetic trapping*); specific RNA-binding proteins are proposed to stabilize the biologically active conformation. (*see also* chaperone machine)

• Herschlag, D. (1995) J. Biol. Chem. **270**, 20871–20874

RNA editing (*see* guide RNA)

RNA fingerprinting (*see* differential PCR display)

RNA fingerprinting arbitrary primer-PCR (*see* differential PCR display)

RNA ligase reverse transcription PCR (RL-RT-PCR) A procedure for amplification of minute amounts of an unknown RNA (or DNA) sequence. The RNA fragment is ligated at both ends to single-stranded palindromic oligodeoxynucleotide that provides a site for subsequent hybridization with a second oligodeoxynucleotide. Reverse transcription uses as a primer the hybridized oligodeoxynucleotide, and then PCR amplification generates a family of polynucleotides that can be treated with a restriction endonuclease to produce large fragments of the cloned RNA and smaller fragments of linker oligonucleotides.

• Wahls, W.P. (1994) PCR Methods Appl. **3**, 272–277

RNA recoding Non-classical reading of the genetic code, e.g. frame-shifting during transcription to find a codon; use of AUG, normally a stop signal, as a coding signal.

RNA regulator An mRNA-like transcript that is not translated but that affects expression of sequence-unrelated DNA.

RNase cleavage A method for detection of single base substitutions in DNA fragments. A labelled RNA probe that spans the target DNA sequence becomes susceptible to RNase A cleavage wherever there is any mismatch in sequences. The cleavage of the probe is detected upon gel electrophoresis.

RNA single-strand conformational polymorphism (rSSCP) A variant of the SSCP technique for detection of mutations in a DNA fragment in which mobility shifts of bands of mutant compared with normal DNA are more pronounced. A phage promoter is added to one of the PCR primers to allow production of an RNA transcript that is then analysed for altered electrophoretic bands on non-denaturing gel electrophoresis.

• Sarkar, G., Yoon, H. and Sommer, S.S. (1992) Nucleic Acids Res. **20**, 871–878

RNA virus A virus with an RNA genome that may be either an mRNA, (+)-RNA, or its complement, (–)-RNA. Class 1 contains (+)-RNA; class 2, (–)-RNA, which is the template for an RNA-dependent RNA polymerase; class 3, double-stranded RNA, in which (+)-RNA is synthesized by an RNA-dependent RNA polymerase; class 4, retrovirus, in which (+)-RNA is a template for an RNA-dependent DNA polymerase (a reverse transcriptase).

RNA world In one scenario for the origin of life, a stage before the evolution of DNA and proteins in which replicative and catalytic functions were carried out exclusively by RNA molecules.

RNP RNA–protein complex. (*see* small nuclear RNA)

ROA Raman optical activity. (*see* vibrational optical activity)

Robison ester Obsolete name for glucose 6-phosphate.

rocket electrophoresis A technique for identification and quantification of a material by electrophoresis into a slab of agarose gel that contains an antibody. The height of the leading edge of the area of precipitation is indicative of the amount of antigen present.

rolling circle replication DNA synthesis that uses circular double-stranded DNA as a template to generate multiple tandem copies of a gene. One strand is nicked and serves as a primer for elongation at the 3'-end, while the unnicked strand serves as a template. Very small synthetic DNA circles may also be used to make new strands with repeating sequences.

Root effect A phenomenon in some teleost fish that allows them to generate O_2 gas in their swim bladders in the face of considerable hydrostatic pressure. The haemoglobin structure and the local secretion of lactic acid create an exaggerated Bohr effect.

• Perutz, M.F. (1996) Nature Struct. Biol. **3**, 211–212

Rossmann fold A supersecondary structure found in some proteins; a motif of β-strands connected by α-helix cross-over elements. (*see also* nucleotide-binding domain).

rotamer distribution (*see* entropy effect)

rotational symmetry (*see* symmetry)

rough endoplasmic reticulum (RER) The endoplasmic reticulum that is studded with ribosomes; the site of synthesis of secretory, intrinsic membrane, Golgi apparatus and ER-resident lysosomal proteins.

RQ (= respiratory quotient)

rRNA Ribosomal RNA; the RNA that forms part of the structure of ribosomes.

rSSCP (= RNA single-strand conformational polymorphism)

R strain (*see* lipopolysaccharide)

RT-PCR (= reverse transcriptase-PCR)

ruffled edge (= lamellipodium)

run-off (*see* nuclear run-off assay)

run-on assay A method for evaluation of the status of replication, transcription or translation in a cell. The assay continues the synthesis of a polymer without reinitiation. (*see also* nuclear run-off assay).

S

S (see R)

SAAB Selected and amplified (protein) binding site oligonucleotide. (see cyclic amplification and selection of targets)

saccharide A sugar or a polymer of sugars linked by glycosidic (acetal or ketal) bonds. (see also mono-; oligo-; poly-)

Sakaguchi reaction A colorimetric reaction for identification and quantification of guanidino groups that involves reaction with α-naphthol and sodium hypochlorite.

Salmonella **test** (= Ames test)

salt bridge An interaction between positive and negative charges on side chains of a protein.

salting in (see salting out)

salting out The addition of a salt, especially ammonium sulphate, to decrease the solubility of a susceptible protein. Some proteins respond by becoming more soluble upon addition of a salt, i.e. *salting in*. (see also globulin)

salvage pathway A metabolic reaction that recovers some of a partially degraded compound for resynthesis, e.g. resynthesis of GTP from guanine recovered from degradation of nucleotides via the hypoxanthine–guanine phosphoribosyltransferase reaction.

sandwich immunoassay A method for quantification of an antigen large enough to have two epitopes. The antigen serves as a bridge between an immobilized (captive) antibody attached to one epitope, and a radioisotope-, fluorophore- or chromogenic enzyme-labelled antibody attached to a second epitope. The amount of immobilized label is directly related to the amount of antigen present.

Sanger method (dideoxynucleotide sequencing) A technique for determination of the base sequence of a homogeneous polynucleotide that acts as a template in a replication system. A small amount of a specific dideoxynucleoside triphosphate is included with the four normal deoxynucleotide triphosphates during replication. Chain termination occurs whenever a dideoxynucleotide is incorporated that generates a labelled polynucleotide whose chain length, evaluated by polyacrylamide-gel electrophoresis, is indicative of the distance from the 5'-end of the primer to the position at which the normal nucleotide occurs. Detection of the electrophoresis bands may be by a label such as ^{32}P or, more recently, by a fluorophore incorporated into the primer (*dye primer sequencing*) or the dideoxynucleotide (*dye terminator sequencing*). In fact, in a variation of the dye terminator method, the labelling of each of the four dideoxynucleotides with a fluorophore that emits at a different wavelength permits the identification of the base that terminates each electrophoresis band within a separated mixture of all four (*single lane sequencing*). (see also Maxam–Gilbert method; plus and minus method)

• Sanger, F., Nicklen, S. and Coulson, A.R. (1977) Proc. Natl. Acad. Sci. U.S.A. **74**, 5463–5467

saponification number A characteristic of an acylglycerol; the number of mg of KOH necessary to neutralize the fatty acids liberated from 1 g of material.

saponin A naturally occurring derivative of a plant steroid bound to a sugar moiety. Those with certain pharmacological properties, e.g. digitonin, are called *cardiac glycosides.*

SAR [= scaffold-associated region (*see* AT queue), or structure–activity relationship]

sarcolemma The excitable plasma membrane of a muscle cell.

sarcoma A malignant tumour of mesodermal origin.

sarcomere The minimal functional unit of skeletal muscle; roughly a cylinder bounded on its ends by Z-lines where the thin filaments are anchored; these extend towards the centre. Thick filaments extend from the centre of the sarcomere, forming the A-region. The thick and thin filaments overlap in two darker-staining zones, which bracket the central H-region.

sarcoplasm The cytoplasm of a skeletal muscle cell.

sarcoplasmic reticulum A flattened membrane-limited compartment that surrounds a myofibril of skeletal muscle and that contains Ca^{2+}, which can be released to stimulate contraction.

sarcosome A mitochondrion of a muscle cell.

satellite DNA Short repetitive DNA sequences that occur mainly at the ends or in the centre of chromosomes and are therefore suspected of serving structural roles; also, polynucleotides that are separable, on the basis of their characteristic density, from the bulk of nuclear DNA, and that have repetitive sequences.

saturation In protein chemistry, the limit at which a reversible binding site is fully occupied by a ligand. In magnetic resonance, saturation is the raising of a paramagnetic compound or chemical group to an activated state by radiofrequency radiation faster than it can spontaneously relax to the ground state; as the high-energy spin state of the molecules is fully populated, higher intensity radiation cannot be absorbed.

saturation mutagenesis (random oligonucleotide mutagenesis) A technique to mutate all bases of a gene. DNA synthesis proceeds from an assortment of oligonucleotide primers synthesized by substitution at each point of the sequence of all three incorrect nucleotides, so that an unbiased assortment of mutants is generated for the span of DNA covered by the primer. Use of other modified primers to cover the length of the gene produces a complete library of mutants.

scaffold-associated region (*see* AT queue)

scaffold protein A structural protein of chromatin or of one of the more complex viruses.

scanning In protein synthesis, the movement of a ribosome along an RNA molecule. Also, the rapid surveying of clones, libraries or DNA sequences for a desired trait. (*see also* genomic mismatch scanning; homologue-scanning mutagenesis; PCR-based differential scanning)

scanning force microscopy (= atomic force microscopy)

Scatchard plot A graphical method for determination of a binding constant and the number of binding sites. The ratio of the amount of ligand (e.g. a hormone) that is bound (e.g. to a receptor) to the amount that is free is plotted against the free amount.

scavenger A compound that can combine with a free radical and thus alter the kinetics or course of free-radical reactions. (*see also* antioxidant)

Schardinger dextrin A product of bacterial degradation of starch; a closed-ring oligosaccharide in which glucosyl monomers are joined by $\alpha 1 \rightarrow 4$ glycosidic bonds.

Schiff base The product formed by the reversible condensation of an aldehyde with an amino group in which the elements of water have been eliminated. It is featured in structures of biochemical interest, notably the reactions of amino acids with pyridoxal phosphate as a coenzyme of various enzymes of amino acid metabolism.

Schiff's reagent (*see* periodic acid–Schiff stain)

schlieren method An optical method for detection of the non-uniform dispersion of a macromolecule in solution, especially during analytical ultracentrifugation or free-zone electrophoresis, by the refractive index of its solution.

scintillation counting A technique for detection and quantification of radioactivity by capture of the emission from a radiolabel by a primary phosphor which, on emission, excites a secondary phosphor and causes it to emit a pulse of detectable visible light. (*see also* excimer)

scintillation proximity assay (SPA) A method for quantification of a molecule on the basis of its containing a specific ligand. The molecule is labelled with a weakly emitting radioisotope and a ligand, e.g. a [^{3}H]glycoprotein. The molecule becomes adsorbed to minute beads that contain a scintillant that is excited by the energy of radioactive decay and a recognition site for the ligand, e.g. concanavalin A. Secondary emission of light from the scintillant is quantified in a scintillation counter. Because a weakly emitting radioisotope is used, only when the radioactive decay occurs in close proximity to the scintillant (i.e. when the molecule is bound to the bead) is a visible emission produced.

scissile bond The bond of a substrate that is subject to enzymic cleavage.

scissors grip A term borrowed from wrestling terminology to describe the interaction of a bZIP protein with the double-standed DNA to which it binds.

scRNA Small cytoplasmic RNA.

SDS gel electrophoresis (= sodium dodecyl sulphate/polyacrylamide-gel electrophoresis)

SDS/PAGE (= sodium dodecyl sulphate/polyacrylamide-gel electrophoresis)

sebum The largely triacylglycerol oily secretion of the sebaceous glands of the skin.

seco- A prefix that signifies a product of the opening of a ring, e.g. a product of A-ring opening, a 2,3-*seco*-steroid, or if the ring-opening is oxidative, a 2,3-*seco*-dioic acid.

secondary-ion mass spectrometry (*see* mass spectrometry)

secondary metabolism Those metabolic pathways, other than those for energy production and for biosynthesis of nucleic acids, proteins, structural components, etc., that are unique to an organism; especially those pathways by which bacteria, moulds and plants synthesize pigments, antibiotics, toxins and other natural products.

secondary structure The regular folding of a protein in repeated patterns, e.g. α-helix, β-pleated sheet, β-turns; also the double-helical structure of a polynucleotide. (*see also* primary structure; quaternary structure; supersecondary structure; tertiary structure)

second genetic code An imprecise term that sometimes refers to the nature of the amino acid residues of a protein which determine its secondary and tertiary structure, and sometimes to the features of a tRNA molecule that make it recognizable by one amino acid synthetase but not by others.

second law of thermodynamics The relationship of free energy (G), enthalpy (H) and entropy (S) when a system changes from one equilibrium state to another: $\Delta G = \Delta H - T\Delta S$, where T is the absolute temperature. The law provides the conceptual basis for redox and group transfer potentials; for example, the Nernst equation or the equilibrium constant, which is $\Delta G = -RT\ln[K_{eq}]$, where R is the gas constant. The law allows prediction of the direction in which reversible reactions may proceed, dependent upon reactant and product concentrations, by quantify-ing and defining free energy for a spontaneous chemical change as a negative value which is independent of the path between the initial and final state. Using transition state theory, the law may be applied to rate constants and their temperature dependencies.

second messenger A chemical signal that is generated inside a cell when a hormone (the first messenger) becomes bound to a surface receptor on the outside; e.g. cyclic AMP.

secretagogue A chemical agent that promotes secretion.

secretory vesicle A membrane-bound organelle near the apical membrane of a cell that contains a protein, neurotransmitter or other substance awaiting a signal for secretion. (*see also* zymogen granule)

sedimentation constant A measure of the rate of sedimentation in an analytical ultracentrifuge; $v/\omega^2 x$, where v is the rate of sedimentation, ω is the angular velocity in radians/s and x is the distance from the axis of rotation; usually expressed in Svedberg units (10^{-13} s).

sedimentation-velocity ultracentrifugation A technique that measures the rate of movement of a dissolved polymer under the force exerted upon it in a centrifuge, to assist in evaluation of molecular mass, axial ratio and other hydrodynamic properties. (*see also* equilibrium sedimentation ultracentrifugation)

segmental flexibility The internal movement possible in some proteins, in which domains show little internal movement but move relative to each other by flexion at the hinges that connect them, e.g.

the wagging movement of the antigen-binding sites of immunoglobulin G towards or away from each other.

selected and amplified (protein) binding site oligonucleotide (*see* cyclic amplification and selection of targets)

selective theory (clonal selection theory) A hypothesis that explains antibody specificity by the existence of a wide variety of antibodies, even before exposure to any particular antigen. (*see also* instructive theory)

SELEX Systematic evolution of ligands by exponential enrichment. (*see* cyclic amplification and selection of targets)

self-assembly The spontaneous aggregation of a complex from its components, the chemistry of which determines the nature of the complex, e.g. the formation of a lipid bilayer from phospholipid molecules.

selfish gene DNA that functions to propagate itself to the detriment of other DNA sequences of the genome; e.g. a DNA that encodes an endonuclease that specifically cleaves an allele that does not contain the nuclease gene, and allows conversion of the heterozygous cell into a homozygous cell that contains the selfish DNA, or a lethal or detrimental gene that persists during the evolution of an organism, perhaps due to its close linkage to essential genes.

- Grivell, L.A. (1992) Curr. Biol. **2**, 450–452; Bull, J.J., Molineux, I.J. and Werren, J.H. (1992) Science **256**, 65

self-splicing The activity of the precursor of mature RNA whereby it catalyses its own splicing. *Group I self-splicing* is assisted by an internal guide sequence near the 5′-end and other sequences that form internal base pairing to bring the structure into a reactive supersecondary structure; reaction proceeds by attack of the 3′-hydroxy group of a guanosine (or guanine nucleotide) on the upstream exon–intron junction, which displaces the 3′-end of the upstream exon; its newly generated 3′-end then attacks the junction between the intron and the downstream exon to generate the spliced exons and the free intron. *Group II self-splicing* differs in that the attacking agent is the 2′-hydroxy group of an adenylate residue of the intron located near its junction with the downstream exon, and intermediates are the upstream exon and the downstream exon still ligated to the intron, whose 5′ guanosine hydroxy group is looped back to the internal adenylate residue and is held in a 2′,5′ phosphodiester bond. The looped intron (a *lariat*) is finally displaced by attack of the 3′-hydroxy group of the upstream exon on the intron–(downstream exon) junction. (*see also* ribozyme)

- Saldanha, R., Mohr, G., Belfort, M. and Lambowitz, A.M. (1993) FASEB J. **7**, 15–24; Belfort, M. (1993) Science **262**, 1009–1010

semi-conservative replication The mode of DNA synthesis that results in each new duplex having one parent polynucleotide strand and one newly synthesized strand. (*see also* Meselson–Stahl experiment)

semi-discontinuous Descriptive of the replication of DNA in which the parent strand that is exposed from its 3′-end towards its 5′-end is used as a template for continuous synthesis of a new strand, but the parent strand that is exposed from its 5′-end towards its 3′-end is used as a template for synthesis of short discontinuous segments of DNA, called *Okazaki fragments*.

semi-permeable Descriptive of a membrane that will permit transport of solvent and small solute molecules but will retain larger solutes. (*see also* dialysis)

semiquinone A free-radical intermediate between a quinol and the more oxidized quinone.

sense strand The strand of double-stranded DNA that contains genetic information; the template or coding strand; often represented as the (+)-strand. (*see also* antisense)

sensitivity In quantitative terms, the responsiveness of a physiological system to a stimulus, such as an allosteric effector, substrate or hormone. Where X is the magnitude of the stimulus and Y the magnitude of the response, sensitivity can be expressed as the dimensionless relative increase of Y per relative increase in X, effectively the slope of a plot of $\log Y$ against $\log X$. Alternatively, sensitivity is expressed as the increase in Y per increase in X, i.e. dY/dX, which has the dimensions of the dependent and independent variables.

• Crabtree, B. and Newsholme, E.A. (1985) Curr. Top. Cell. Regul. **25**, 21–76

sequenator A device to sequentially degrade a protein to determine its primary structure.

sequence The ordered linear array of amino acid residues in a protein, of bases in a polynucleotide, or of glycosides in a complex polysaccharide; also, as a verb,

to determine such a sequence, e.g. to sequence a protein.

sequence insertion (*see* gene targeting)

sequence replacement (*see* gene targeting)

sequence-tagged connector (STC) An approximately 500 bp sequence at each end of a bacterial artificial chromosome (BAC). If the human genome is cleaved into 300 000 fragments, of an average length of 150 kb, there would be 600 000 STCs, representing about 10% of the entire genome sequence. STCs may be used as markers to orient within the genome, sequences of randomly selected fragments. (*see also* shotgun sequencing)

• Venter, J.C., Smith, H.O. and Hood, L. (1996) Nature (London) **381**, 364–366

sequence-tagged site (STS) A landmark in the base sequence of a genome used for placing one of many partial sequences in correct relation to others, in which 200–500 bp sequences are determined and stored in a database; especially as a strategy for co-ordination of the efforts of many laboratories in sequencing the human genome. (*see also* contig map; linkage map; physical map; restriction map)

• Marx, J. (1995) Science **270**, 1919–1920

sequencing The determination of the base sequence of a homogeneous DNA fragment, the amino acid sequence of a protein, or the glycoside sequence of a complex polysaccharide.

sequencing by hybridization A proposed approach to rapid sequencing of genomic DNA. The key tool for analysis is a *superchip*, a surface over which are attached at minute yet identifiable positions unlabelled oligonucleotide probes.

For example, if pentanucleotides are used, there will be 625 (5^4) rows of unique pentanucleotides. The entire surface of the superchip is exposed to an amplified sample of the test DNA plus a ligase. Each unlabelled probe will hybridize to the DNA where there is a complementary sequence. The superchip is then overlaid by another chip that contains a similar array of rows of labelled oligonucleotides, but with the rows running across the rows of unlabelled probes. Where the labelled and unlabelled probes are complementary to adjacent sequences (in the example, a decanucleotide sequence), the ligase can join them and thus fix the label to the superchip. A robotic reader will identify the positions on the superchip that have become labelled and that therefore indicate the presence of a particular (decanucleotide) sequence. The overlap of the identified sequences will allow reconstruction of the entire sequence of the test DNA.

sequential (*see* enzyme mechanism)

sequential feedback A control mechanism for a branched metabolic pathway. The end product of each branch inhibits the first enzymic step of its own branch, which results in an accumulation of the last metabolite of the common pathway. This causes inhibition of the first step of the common pathway and results in inhibition of the common pathway if both end products are present in excess.

sequential model (= Koshland model)

serotype An immunologically defined variant of a micro-organism; one that bears a unique epitope on its surface.

serine proteinase A type of peptidase that has at its active site a serine residue. (*see also* aspartate proteinase; cysteine proteinase; metalloproteinase)

serpin Serine proteinase inhibitor; one of a group of homologous proteins, e.g. α_1-antiproteinase.

• Potempa, J., Korzus, E. and Travis, J. (1994) J. Biol. Chem. **269**, 15957–15960

serum The liquid phase that remains after blood has clotted. (*see also* plasma)

sesquiterpene (*see* terpene)

seven-helix motif A characteristic of several receptor proteins, that zig-zag seven times as α-helices across the plasma membrane. (*see also* transmembrane domain)

sex pilus In bacterial conjugation, an appendage of a male bacterium by which it attaches to a female bacterium, preparatory to the transfer of DNA from male to female.

shape library An assortment of oligonucleotides, often of random sequence, some of which (or some of the encoded proteins of which) can bind to secondary structures of other oligonucleotides or proteins. The common features of the selected polymers are termed the *recognition consensus sequence* (*RCS*).

• Kenan, D.J., Tsai D.E. and Keene, J.D. (1994) Trends Biochem. Sci. **19**, 57–64

shikimic acid metabolite (C_6C_3 metabolite) A derivative of shikimic acid with an aromatic 6-carbon ring and an attached 3-carbon side chain, e.g. cinnamic acid, phenylalanine.

Shine–Delgarno sequence A polynucleotide sequence of prokaryotic mRNA that is complementary to, and therefore base-pairs with, a sequence at the 3'-end of bacterial 16 S rRNA, which function to position the initiator codon at the peptidyl site of the ribosome.

short-chain fatty acid (*see* medium-chain fatty acid)

short-column method An analytical ultracentrifugation technique for measurement of the molecular mass of a protein; requires relatively brief periods of time because it uses very short columns of liquids since the distance from meniscus to base is 3 mm or less.

short tandem repeat (STR) A 2–6 bp sequence in genomic DNA that is repeated a number of times that is characteristic of an individual. (*see also* variable number tandem repeat)

shotgun sequencing An approach to determination of the sequence of a large, randomly selected, piece of genomic DNA. A cosmid that contains the DNA is sonicated into smaller fragments of 1.0–1.5 kb, which are ligated into a vector (e.g. M13) containing a universal primer, and each fragment is sequenced. In order to obtain the complete linear structure of a 40 kb cosmid, 800–1000 fragments must be sequenced. More recently, bacterial artificial chromosomes (BACs) have been employed, allowing up to 350 kb fragments to be cloned for sequence analysis.

• Venter, J.C., Smith, H.O. and Hood, L. (1996) Nature (London) **381**, 364–366

shrinkage temperature The temperature at which collagen fibres contract; analogous to the melting temperature of a protein.

shuffling Recruitment of the exons that code for a functional domain of one protein in the evolution of a gene for a new protein, to produce hybrids with recognizable motifs, such as kringles and metal-binding fingers. A consequence of this hypothesis is the corollary that introns are as old as structural genes; this is upheld by the intron-early school but opposed by the intron-late school, who point to the absence of introns in prokaryotes and some primitive eukaryotes.

• Gibbons, A. (1992) Science **257**, 30–31; Stoltzfus, A., Spencer, D.F., Zuker, M., et al. (1994) Science **265**, 202–207; Hurst, L.D. (1994) Nature (London) **371**, 381–382; Mattick, J.S. (1994) Curr. Opin. Genet. Dev. **4**, 823–831

shunt Usually, the pentose phosphate pathway.

shuttle A mechanism for transport of metabolites or chemical groups across the mitochondrial membrane, e.g. the electron shuttle that transports glycerophosphate into, and dihydroxyacetone phosphate out of, the mitochondrion, which permits oxidation of NADH in the cytosol and reduction of FAD inside the mitochondrion.

shuttle vector A plasmid that has both bacterial and eukaryotic origins of replication and so can propagate in either kind of cell; useful as a form of recombinant DNA for growth in a bacterium and subsequent transfer into eukaryotic cells for expression.

side chain The moiety of an amino acid residue in a protein, or of a free amino acid, that is attached to the α-carbon and is unique to each amino acid, e.g. the isopropyl group of valine, the benzyl group of phenylalanine. The group attached to C-17 of sterols, mineralocorticoids and bile acids is also called a side chain.

siderophore A low-molecular-mass compound that binds to ferric ions and facilitates their absorption by the micro-organisms that produce them; e.g. enterobactin, the catechol derivative of some bacteria, and ferrichrome, the hydroxamate compound of some fungi.

• Neilands, J.B. (1995) J. Biol. Chem. **270**, 26723–26726

side shelf experiment An experimental approach to discovery of a metabolic precursor or regulator, in which candidates for activity are randomly selected according to availability rather than by intelligent design.

sigma (σ) subunit The subunit of RNA polymerase that recognizes transcription initiation sites on double-stranded DNA.

sigmoidal S-shaped; descriptive of the appearance of a graph that shows dependence of rate on substrate concentration for a multimeric enzyme that is accelerated by binding of more than one substrate molecule; also descriptive of nonsimple binding of ligands to a nonenzymic protein, e.g. oxygen binding to haemoglobin.

signal molecule A neurotransmitter, hormone, second messenger or other regulatory molecule.

signal patch An area of a protein surface that is recognized by the sorting mechanism of a cell and allows the protein to be directed to its appropriate locus within the cell.

signal recognition particle (SRP) In the synthesis of secretory, endoplasmic reticulum (ER), Golgi apparatus, lysosomal and integral membrane proteins, a ribonucleoprotein that recognizes a signal sequence and prevents further translation until the docking protein (SRP receptor) of the ER can direct the growing polypeptide across the membrane of the ER. (*see also* ribophorin)

signal sequence (leader sequence) The N-terminal portion of a secretory or membrane protein that assists it across the membrane of the rough endoplasmic reticulum, where it is synthesized, but is cleaved from the protein even before the synthesis of the protein is complete. Mitochondrial and chloroplast proteins that are synthesized on ribosomes also have signal sequences that target them to their organelle; these may be N-terminal or internal, and may or may not be removed by proteolysis.

silencer (*see* origin recognition complex)

silencing The genetically programmed repression of transcription, e.g. due to a DNA-binding protein which interacts with a regulatory element that controls the silenced gene. (*see also* origin recognition complex)

• Schoenherr, C.J. and Anderson, D.J. (1995) Curr. Opin. Neurobiol. **5**, 556–571

silent mutation A mutation that causes no functional change in the gene product. (*see also* transparent mutation)

simple sequence DNA The highly repetitive non-coding DNA of a satellite that serves a structural role in mitosis; used for genetic characterization by its polymorphism (*simple sequence length polymorphism, SSLP*).

simplified Descriptive of a protein whose amino acid sequence has been modified to minimize its complexity, e.g. by substitution in many sequence positions with only a few selected amino acids.

SIMS Secondary-ion bombardment mass spectrometry. (*see* mass spectrometry)

simulated annealing An energy minimization technique to refine a model structure, e.g. an X-ray crystallographic or nuclear magnetic resonance model, by simulation of heating it to a high temperature to break all covalent and non-covalent bonds and overcome local energy minima, recalculation of the positions of each atom, then slowly cooling it to allow bonds to re-form between neighbouring atoms.

SINE (= small interspersed repeat element)

Singer–Nicolson model (= fluid mosaic model)

single-domain antibody (dAB) The variable domain of an immunoglobulin heavy (V_H) or light (V_L) chain, each of which can have considerable affinity for an antigen; produced in bacteria by biotechnological methods, whereby the newly synthesized protein can pass freely through the cell wall into the medium.

single locus DNA profiling The use, e.g. in forensic science, of a single variable number tandem repeat to characterize an individual's DNA.

single-lane sequencing (*see* fluorescence-based DNA analysis technology)

single-pass sequencing (= single-lane sequencing; *see* fluorescence-based DNA analysis technology)

single-strand conformational polymorphism (SSCP) A technique for detection of mutations in a defined DNA sequence. A labelled PCR-amplified segment of DNA is heat-denatured then quenched at 0°C, and the resultant single-stranded polynucleotides are electrophoretically separated on non-denaturing gels. Intrachain base pairing results in a limited number of conformers stabilized by intrachain loops, and mutated DNA shows on electrophoresis an altered assortment of such conformers. (*see also* restriction endonuclease fingerprinting)

singlet state An excited electronic state in which an electron is raised to a higher energy level without reversal of its spin state, and thus is easily able to fall back to its ground state, often with a fluorescent emission. (*see also* triplet state)

singlet strand An unpaired polynucleotide chain.

single-well hydrogen bond (*see* hydrogen bond)

site-directed mutagenesis A technique to alter deliberately a structural gene to insert a desired point mutation. A synthetic polynucleotide that contains the genetic alteration is used as an imperfectly fitting primer in replication and the completed (–)-strand incorporates the mutation.

site-specific mutagenesis (= site-directed mutagenesis)

slab gel A thin sheet that acts as a support for electrophoresis, often composed of polyacrylamide or agarose. (*see also* disc-gel electrophoresis)

sliding filament model A conceptualization of muscle contraction in which thick filaments in the centre of a sarcomere are surrounded by thin filaments that are attached to the ends of the sarcomere and extend towards its centre. Stimulation drives the overlap of the thick and thin filaments and thus pulls the ends of the sarcomere towards the centre.

slow-twitch muscle (= red muscle)

small interspersed repeat element (SINE) A group of recurring species-specific polynucleotide sequences in the human genome, e.g. an *Alu* sequence of 0.3 kb present in approximately 10^6 copies. (*see also* long interspersed repeat element)

small molecule library A collection of related structures synthesized for screening for a desired property. The library may by synthesized by parallel, split or spatially addressable synthesis. *Parallel synthesis* is a single-batch method that uses a mixture of reagents at each step of the synthesis to generate the diversity of products. *Split synthesis* requires a solid-phase support to allow separation of products at each synthetic step. For example, a polymer may be assembled by coupling monomers A, B and C to beads; the separate batches are filtered and combined, then divided again into thirds for coupling with monomers D, E and F. After n cycles, there will be 3^n sequences present. *Spatially addressable synthesis* localizes the reactions on specific locations on a chip, a solid surface ruled into a grid, each locus on which receives a unique sequence of reagents. The products of each synthetic method are exposed to a selective procedure. The position on a chip directly identitfies the structure assembled there. If a binding property is sought, localization to a ligand on a solid-phase surface will assist selection from a library of dissolved structures. Coding of beads during their synthesis allows their identification after selection; such coding may be the parallel assembly of a protein or nucleic acid during each step of the synthesis, polymers of which are easily sequenced by available technologies.

• Borman, S. (1996) Chem. Eng. News **74 (7)**, 28–54

small nuclear RNA (snRNA) A category of polynucleotides, usually from 70 to 500 nucleotides in length. As complexes with proteins (*snrps*) they have several functions, among them the processing of pre-mRNA, either directly as part of a spliceosome or indirectly by hybridizing with the pre-mRNA and altering its architecture, i.e. by altering its secondary structure, its tertiary and quaternary structure is also altered. Some snrps are encoded by their own genes, others by the introns of a host gene which itself has some ribosome-related function. (*see also* spliceosome)

smooth endoplasmic reticulum (= endoplasmic reticulum; *see also* rough endoplasmic reticulum)

SMP (= submitochondrial particle)

snap-back sequence Half the base-pairing sequence of a hairpin turn of a tRNA, which can easily rebind to its complementary sequence after denaturation.

snRNA (= small nuclear RNA)

snrp Small nuclear RNA–protein complex. (*see* small nuclear RNA)

S₁ nuclease mapping A technique for identification of nucleotides at the 5'-end of an mRNA which, because it retains at that end its original triphosphate moiety, may be specifically labelled for sequencing by the Maxam–Gilbert technique. A restriction fragment that is complementary to the 5'-end of the mRNA protects it against S_1 nuclease digestion, which is specific for single-stranded RNA and DNA; the only remaining polynucleotides therefore are the 5'-end of the mRNA and its unlabelled complement.

snuRNA Small nucleolar RNA.

soap A sodium, potassium or other soluble salt of a fatty acid.

sodium dodecyl sulphate/polyacrylamide-gel electrophoresis (SDS/PAGE) A technique to dissociate multimeric proteins and to separate all proteins according to their apparent molecular masses. Proteins are treated with sodium dodecyl sulphate (SDS), an anionic detergent, and separated by electrophoresis on polyacrylamide gels that incorporate SDS. The detergent adsorbs on to the proteins in proportion to their mass and dominates their electric charges. Also known as denaturing gel electrophoresis or SDS gel electrophoresis. (*see also* native gel)

• Laemmli, U.K. (1970) Nature (London) **227**, 680–685

solenoidal Descriptive of one form of slightly unwound supercoiled DNA in which the double helix lies in coils that are stacked one upon another in a very condensed arrangement. (*see also* solenoidal model)

solenoidal model A representation of the organization of chromatin in which adjacent nucleosomes in the linear DNA sequence are wound into a 36 nm diameter helix that contains six nucleosomes per turn.

sol-gel immobilization A method for making an enzyme encapsulated in a porous silicate matrix. An anhydrous preparation of tetramethylorthosilicate begins to hydrolyse as soon as it is in contact with an aqueous enzyme solution. The newly formed hydroxy groups of the silicate condense with themselves and with hydroxymethyl groups to form a water-sol phase, and then a gel that entraps the enzyme. The result is a glass-like transparent solid that may have uses in biotechnology, e.g. in a biosensor.

solid-phase sequencing A set of methods for sequencing of proteins and synthesis of proteins and nucleic acids, in which the subject molecule is attached to a solid support to aid in its separation from reagents during chemical reactions. The reagents may be in a liquid or a gas phase. In one procedure, the reaction takes place at the solid–liquid interface of a *spinning cup* that holds the immobilized polymer and the reagent solution.

soluble RNA (= tRNA)

solvation substitution (*see* mobile barrier)

solvent-accessible surface A feature of a macromolecule defined by a molecular modelling technique, e.g. that of Connolly (a *Connolly surface*); the parts of the structure that are exposed to the solvent molecules are identified by rolling a sphere the size of a water molecule over its surface.

solvent isotope effect The effect on an enzyme's kinetics of inclusion of ^2H_2O; often expressed as ^Dk, the ratio of the reaction rate constant in H_2O to the rate constant in ^2H_2O. An extension of this approach is a *proton inventory*, in which the rate constant of each phase of an enzymic transformation is determined as a function of the mole fraction of ^2H_2O, in order to determine the number of protons transferred in each discrete step of the catalytic cycle.

solvophobic effect (= hydrophobic effect)

somatic cell A non-gamete.

somatic cell hybrid mapping (*see* mapping; radiation-induced hybrid mapping)

somatic gene therapy (*see* gene therapy)

somatic hybrid The fusion product of two different types of non-gamete cells.

somatic mutation A mutation in a non-gamete cell; therefore one that will not be passed to progeny but only to daughter cells generated by mitosis.

sonication A technique to disrupt cell structures by subjecting them to high-frequency sound waves.

Soret band A spectrophotometric characteristic of porphyrins; a strong absorption band near 400 nm.

sorting The translocation to their appropriate loci within a cell of newly synthesized or endocytosed proteins and nucleic acids.

SOS repair An error-prone DNA repair process induced in response to ultraviolet-light-induced mutagenesis. In the face of pyrimidine dimerization, it allows for continued DNA resynthesis with relaxation of the strict requirement for base pairing.

• Walker, G.C. (1995) Trends Biochem. Sci. **20**, 416–420

Southern blotting A technique for detection of specific DNA fragments. Restriction fragments are separated by polyacrylamide-gel electrophoresis, blotted on to a plastic sheet (e.g. of nitrocellulose) and detected by hybridization with a radiolabelled probe followed by autoradiography. (*see also* electroblotting; northern blotting; south-western blotting; western blotting)

south-western blotting A technique for identification of DNA-binding proteins and regions of DNA that bind proteins which combines aspects of Southern and western blotting techniques. A nuclear extract is separated by sodium dodecyl sulphate/polyacrylamide-gel electrophoresis, renatured and blotted on to nitrocellulose, then exposed to labelled DNA probes. Photographic film laid against the gel is exposed so as to identify which protein bands have bound DNA probes, and then the DNA is extracted from them. (*see also* Southern blotting; western blotting)

SPA (= scintillation proximity assay)

spacer arm In affinity chromatography, the chain of carbon and/or other atoms that positions a functional group away from the solid matrix to which it is covalently bound and makes it more available to a ligand and less restricted by steric hindrance by the matrix.

spacer DNA DNA that does not yet have a recognized function. (*see also* junk DNA)

spare receptors The phenomenon of a hormone-responsive cell having more receptors capable of binding the hormone than are needed to effect a maximal response; possibly an adaptation that makes the cell responsive to the rate at which the hormone can reach the cell surface (the *collisional limit*).

spatially addressable synthesis (*see* small molecule library)

SPCR (= synthetic peptide combinatorial library)

species A group of organisms that breed among themselves to produce viable offspring.

specific acid In chemistry, a hydronium ion that can participate in catalysis. (*see also* general acid; specific base)

specific activity A measure of function per unit, e.g. enzyme activity per mg of protein, or radioactivity per weight (or per volume, or per mole) (more correctly known as specific radioactivity).

specific base In chemistry, a hydroxyl ion that can participate in catalysis. (*see also* general base; specific acid)

specific dynamic action (= thermic effect of food)

specificity constant (k_{cat}/K_m) The apparent second-order rate constant for enzyme action under conditions where an enzyme's binding site is largely unoccupied by substrate ([S] « K_m); equivalent to the slope of the (pseudo) first-order region of a plot of enzymic rate against substrate concentration.

specificity subsite A part of the active site of a peptidase. Amino acid residues adjacent to the scissile bond are designated P_1, P_2, P_3, etc. on the N-terminal side, and P_1', P_2', P_3', etc. on the C-terminal side; P_1–P_1' is the scissile bond. Peptidase subsites that bind these are designated S_1, S_1', S_2, S_2', etc.

specific linking difference (= superhelical density)

specific radioactivity A measure of the degree of labelling with a radioactive tracer; sometimes referred to as specific activity; expressed as radioactivity per unit mass, e.g. μCi/mmol, c.p.m./mg.

spermatozoon A male gamete, characterized in part by an acrosome and a flagellum.

S phase (*see* cell cycle)

spheroplast A plant cell that has been incompletely stripped of its cell wall. (*see also* protoplast)

spindle The microtubule structure that radiates from each pole of a cell to the chromatids during mitosis or meiosis.

spin imaging (= zeugmatography)

spin label A chemical functional group with an unpaired electron attached to a compound or macromolecule, e.g. a nitroxide group; detectable and characterizable by its electron spin resonance spectrum.

spin lattice The magnetic environment that surrounds the nucleus or electron under observation in nuclear magnetic resonance or electron spin resonance studies.

spinning cup (*see* solid-phase sequencing)

spliced protein (*see* protein splicing)

spliceosome A complex of (small nuclear RNA)–protein complexes and other proteins that assemble on a pre-mRNA and catalyse the excision of an intron in a process mechanistically similar to group II self-splicing. It acts presumably by forcing the intron into a loop and bridging the pre-mRNA at its splicing sites.

• Wise, J.A. (1993) Science **262**, 1978–1979

3'-splice site (= acceptor splice site)

5'-splice site (= donor splice site)

splicing The process by which an intron is excised and exons are religated in the post-transcriptional modification of RNA; also the excision of an intein from a precursor protein. (*see also* protein splicing; self-splicing; spliceosome)

splinkerette PCR (*see* vectorette PCR)

split gene A gene whose coding sequence is interrupted by introns.

split synthesis (*see* small molecule library)

splitting The generation of multiple signals in nuclear magnetic resonance and electron spin resonance spectroscopy due to the interaction of nuclei or electrons coupled through space. (*see also* hyperfine splitting; Zeeman splitting)

SPR Surface plasmon resonance. (*see* biomolecular interaction analysis)

spreading factor An agent that permits the diffusion of foreign materials through tissue, e.g. hyaluronidase, which degrades the extracellular matrix.

squelching (*see trans*-acting)

***src* gene** A polynucleotide that in one form, c-*src*, is a normal cellular gene, and in a similar form, v-*src*, is an oncogene of avian sarcoma virus.

src protein A product of the *src* gene; acts to phosphorylate tyrosine residues of some proteins.

sRNA Soluble RNA. (= tRNA)

SRP (= signal recognition particle)

SRP receptor (= docking protein; *see* signal recognition particle)

SSCP (= single-strand conformational polymorphism)

ssDNA Single-stranded DNA.

SSLP Simple sequence length polymorphism. (*see* simple sequence DNA)

ssRNA Single-stranded RNA.

S strain (*see* lipopolysaccharide)

staggered ended (= sticky ended)

standard state For solutions in general, the conditions under which thermodynamic properties of a solute are defined, namely 1 M. The properties of the solution are those extrapolated from very dilute solutions. In biochemistry the standard state is defined for $[H^+] = 10^{-7}$ M (pH 7), and thermodynamic properties of all other ionizing species are defined for a total of all protonated and deprotonated species of 1 M.

start signal In translation, the initiation codon (AUG); in transcription, the site of initiation of RNA synthesis.

state 1, state 2, etc. (*see* respiratory state)

STC (= sequence-tagged connector)

steady state In kinetics, the maintenance of the concentration of an intermediate by its formation from precursors at the same rate as its conversion into products.

stefin (*see* cystatin)

stem cell (blast cell) A relatively undifferentiated cell that can develop into a more specialized cell.

stereoelectronic control (*see* entropy effect)

stereoisomer One of two or more compounds that differ only in their orientation at one or more asymmetrical centres. (*see also* diastereoisomer; enantiomer; epimer)

stereopopulation control (*see* entropy effect)

stereospecific transfer The stereospecificity of an enzyme that catalyses either the abstraction of an atom or group from a prochiral centre, or the addition of an atom or group to form one.

steric hindrance The constraint on a reaction or conformational change that is due to crowding of atoms within the van der Waals radii of other atoms.

steroid A derivative of perhydrocyclopentanophenanthrene; a more oxygenated product of cholesterol, with the C-17 side chain shortened or removed. Many steroids are hormones, e.g. cortisone, progesterone, oestradiol. (*see also* sterol)

steroid receptor (*see* nuclear receptor)

sterol A crystalline alcohol that incorporates a characteristic perhydrocyclopentanophenanthrene ring system and a branched hydrocarbon side chain, e.g. cholesterol, ergosterol. (*see also* steroid)

sticky ended Descriptive of the structure of a double-stranded polynucleotide in which one strand extends further from the end than the other to create a single-stranded 'tail'; especially the product of cleavage by a restriction endonuclease at a palindromic sequence. Also known as cohesive ended or staggered ended. (*see also* blunt ended)

stochastic Determined by the laws of chance and probability.

Stokes' Law A theoretically derived relationship between the viscosity coefficient, η, the frictional coefficient, f, and the radius of a macromolecule (the Stokes radius) r: $f = 6\pi\eta r$. This calculated f may be compared with that determined experimentally from the relationship between the diffusion coefficient, D, the Boltzmann constant, k, and the absolute temperature, T: $D = kT/f$. The deviation (usually an increase) of the measured f from the calculated f is a measure of the solvation of the macromolecule or its deviation from a spherical shape.

Stokes radius On the assumption that it is a sphere, the apparent radius of a macromolecule in solution, as determined from its hydrodynamic behaviour, i.e. intrinsic viscosity, diffusion or sedimentation coefficient, or behaviour in gel filtration chromatography. (*see also* Stokes' Law)

stopped-flow A rapid mixing technique for observation of fast reactions, usually in the millisecond range. Two syringes holding solutions that must be mixed to initiate the reaction are emptied by a single piston that forces the contents through a mixing chamber into an observation chamber where spectral properties are observed and stored in an oscilloscope for subsequent analysis.

stop-transfer sequence A polypeptide sequence of a nascent membrane protein that prevents its translocation into a transfer vesicle and permits its insertion into a membrane.

STR (= short tandem repeat)

(−)-strand Minus-strand (*see* plus-strand)

(+)-strand (= plus-strand)

strand displacement-assimilation model (*see* forced copy-choice model)

streak The placement of a bacterial culture on solid media, especially with an inoculation loop or pipette, with the intention of covering the surface in a deliberate pattern.

stress-fibre An actomyosin-containing filament of a non-muscle cell. Between mitoses (interphase) the fibres are anchored at membrane attachment sites and are involved in extension of lamellipodia and in cell motility.

striated muscle Muscle that on microscopic examination shows a banding pattern that is due to the regular spacing of areas of overlap of thick and thin filaments within each sarcomere; includes skeletal muscle and cardiac muscle.

stringency During nucleic acid hybridization or reassociation, the strictness with which the Watson–Crick base-pairing is required under specified conditions of temperature, pH, salt concentration, etc. Conditions of high stringency require all bases of one polynucleotide to be paired with complementary bases on the other; conditions of low stringency allow some bases to be unpaired.

stringent bacteria (*see* magic spot nucleotides)

stroma The supporting connective tissue of an organ. (*see also* parenchyma)

structural profile An approach to prediction of similarities in protein conformations by comparison of amino acid sequences. A two- or three-dimensional plot of one or two features, e.g. optimal matching hydrophobicity, charge, likelihood of occurrence in α-helix or β-sheet, against their sequence number allows visual scanning to identify similarities of profiles and, presumably, similarities of conformations. (*see also* hydrophobic cluster analysis)

structure–activity relationship (SAR) A data set that is the product of screening a potential drug to indicate which of its atoms contribute to its effect. (*see also* AT queue)

STS (= sequence-tagged site)

subdomain model (*see* framework model)

submitochondrial particle (SMP) A preparation obtained from mitochondria that is capable of electron transport.

subsite (= specificity subsite)

substrate A reactant in an enzymic reaction.

substrate anchoring (*see* entropy effect)

substrate channelling The sequential action of a multienzyme complex upon a substrate that prevents the dissociation of intermediate products.

substrate cycle (= futile cycle)

substrate inhibition The inhibition of an enzymic reaction at high substrate levels; usually due to a second, inhibitory and lower-affinity binding site for the substrate in addition to the catalytic site.

substrate-level phosphorylation The metabolic synthesis of ATP that does not require the electron transport chain, e.g. the glyceraldehyde-3-phosphate dehydrogenase reaction.

substrate phage A method for identification of the substrate specificity of a proteinase by its ability to select a phage that expresses a cleaveable polypeptide, which is one of a library of random amino acid sequences fused to a protein that tethers the phage to a solid support. Selection of the phage is by cleavage of the polypeptide. Several cycles of binding, proteolysis and phage propagation select the phage that expresses the scissile bond.

- Matthews, D.J. and Wells, J.A. (1993) Science **260**, 1113–1117

subtractive DNA cloning A difference cloning method; a technique for comparison of two closely related cell types (e.g. a cell with a deletion mutation and the normal cell), and isolation from one (the *tester*) of DNA that is absent from the other (the *driver*). Tester DNA is cleaved with a restriction nuclease that will allow its subsequent insertion into a cloning vector; an excess of driver DNA is randomly sheared. The two preparations are mixed, melted and annealed to form, among other species, duplexes composed of strands of the different sources (*inter-resource duplex, IRD*). The fraction of the annealed product, both strands of which were cleaved at restriction sites and which therefore is capable of being cloned, will be enriched in those sequences for which there was no competing randomly sheared driver DNA, i.e. those that are unique to the tester DNA, especially when annealing is terminated before an equilibrium is established (*kinetic enrichment*). The unique DNA sequences will therefore form a high proportion of the successfully cloned annealed DNA product. (*see also* magnet-assisted subtractive technique; subtractive hybridization)

subtractive hybridization A variation of subtractive DNA cloning in which the restriction-nuclease-cleaved tester DNA is biotinylated before annealing with an excess of randomly sheared driver DNA. Annealing is halted at about 90% completion and the single-stranded DNA is separated from double-stranded DNA.

As the unique tester fragments are present at lowest concentration, they will be slowest to find complementary strands and anneal and will, therefore, be enriched in the remaining single-stranded DNA fraction. They are isolated on the basis of biotin's affinity for immobilized avidin and subjected to subsequent cycles of purification.

- Wieland, I., Bolger, B., Asouline, G. and Wigler, M. (1990) Proc. Natl. Acad. Sci. U.S.A. **87**, 2720–2724

subunit One of the identical or non-identical protein molecules that make up a multimeric protein; also one of the ribonucleoprotein complexes that make up the ribosome. (*see also* heavy subunit; light subunit; protomer)

sucrose-gradient centrifugation A technique for characterization or preparation of subcellular particles. A swinging-bucket-type centrifuge tube is filled with a sucrose gradient, the bottom of which is most dense and the top least dense. A suspension of the particles is layered over the top of the solution, and centrifugation separates the particles within the gradient according to their density. At the termination of centrifugation the bottom of the tube is pierced and the sucrose, with particles of equal density, drips into a series of receiving tubes.

sugar A simple carbohydrate that consists of a chain of carbon atoms, one of which is a carbonyl carbon (aldehyde or ketone), while the others bear hydroxy groups to give it the basic formula $(CH_2O)_n$, sometimes condensed into a hemiacetal or hemiketal; sugar derivatives include oxidation and reduction products, phosphate and sulphate esters, and amino derivatives.

suicide inhibitor (*see* mechanism-based inhibitor)

sulphatide A sulphate ester of a ceramide or related compound, e.g. galactosyl-3-sulphate ceramide, ceramide-dihexoside sulphate.

sulphur cycle The passage of sulphur from organic to inorganic forms, from substituents of cells to minerals, via the following steps: (1) uptake of sulphate from soil by organisms and conversion into organic sulphur, or reduction to the gas dimethylsulphide (e.g. by algae and cyanobacteria), (2) photochemical oxidation to methanesulphonic acid in the atmosphere, (3) return to the soil as it falls to earth in rain and snow, and (4) oxidation to sulphate by soil bacteria.

• Baker, S.C., Kelly, D.P. and Murrell, J.C. (1991) Nature (London) **350**, 627–628

supercoiling The torsional stress placed on double-stranded DNA, maintained for example by interaction with proteins or by DNA that is circular, and accommodated by a twist imposed on the duplex. A left-handed supercoil favours unwinding of the double helix; a right-handed supercoil favours tighter winding. (*see also* superhelical density; topology)

• Schlick, T. (1995) Curr. Opin. Struct. Biol. **5**, 245–262

superfamily A group of genes that are related by evolutionary divergence to an ancestral gene, but that code for products with different functions. (*see also* multigene family)

superhelical density (specific linking difference) A quantitative measure of the degree of supercoiling of a DNA molecule. The value, σ or λ, equals $(L-L_o)/L_o$, where L_o is the linking number of the relaxed double-stranded DNA molecule and L is the linking number for the supercoiled DNA. A negative value signifies a left-handed supercoil.

superhelix The twisted axis of a fibrous polymer, such as double-stranded DNA, myosin tails or collagen. (*see also* coiled coil)

superinduction The additional increase in the rate of synthesis of a protein caused by interference with the cellular apparatus for protein or mRNA synthesis that follows an increase in transcription of the gene. The mechanism varies with circumstances, but results in an increase in the steady-state level of the specific mRNA.

supernatant fluid The unsedimented portion that remains after centrifugation. (*see also* pellet)

supersecondary structure The arrangement of elements of secondary structures in a protein (a motif), e.g. β-barrel, or in a nucleic acid, e.g. cloverleaf. (*see also* primary structure; quaternary structure; secondary structure; tertiary structure)

suppressor mutation A mutation that reverses the effect of an earlier mutation, e.g. a mutation in a gene for a tRNA that permits it to read and override an amber mutation.

suppressor T-cell (*see* T-cell)

surface plasmon resonance (*see* biomolecular interaction analysis)

surfactant (= detergent)

Svedberg unit (*see* sedimentation constant)

switching During development, the turning off of one gene and the turning on of a related one; e.g. during fetal development, the change from synthesis of γ-globin to synthesis of β-globin.

syllabicity (*see* zinc finger)

symmetry One of the geometrical properties of a chemical or any other structure. *Reflectional symmetry* is the property of a molecule or object that permits it to be superimposed on its mirror image; *rotational symmetry* is a property that allows it to be rotated so that, in the new orientation, like groups exchange positions and the rotated orientation is indistinguishable from the original. (*see also* chirality; *meso*-carbon; Ogston hypothesis)

symport A transport mechanism that simultaneously drives two different compounds or ions in the same direction across a membrane. (*see also* antiport; mobile barrier; mobile carrier; uniport)

synapomorphy A characteristic common to a group originating from a common ancestor.

synapse The junction between two interacting neurons.

synaptic cleft The narrow gap between two interacting neurons.

synaptic vesicle A secretory vesicle that stores a neurotransmitter at the presynaptic membrane, i.e. on the afferent side of a synaptic cleft.

synaptosome A vesicle formed by artificially pinching off and resealing the presynaptic ending of a neuron, i.e. the afferent portion of a synapse.

synchronous growth The co-ordinated passage through the cell cycle of a large mass of cells; sometimes achieved by experimental starvation of the cells to bring them all to the start of S phase, then relieving the inhibition by refeeding.

syn **conformation** (*see* Z-DNA)

syndecan (*see* proteoglycan)

synkinesis (*see* biosynthesis)

synkinon (*see* biosynthesis)

synonymous In molecular genetics, descriptive of a nucleotide substitution in a structural gene that does not result in an amino acid substitution; contrasted with *non-synonymous*, descriptive of base substitutions that do result in amino acid substitutions.

synthase A lyase (acting in the reverse of the reaction) that adds one substrate across the double bond of another, e.g. δ-aminolaevulinic acid synthase, hydroxymethyl glutaryl-CoA synthase. (*see also* synthetase)

synthetase An enzyme, regardless of its mechanism, that couples two substrates by a carbon–carbon, carbon–oxygen, carbon–sulphur or carbon–nitrogen bond that is driven by the hydrolysis of a phosphoanhydride bond, e.g. amino acyl-tRNA synthetase, fatty acyl-CoA synthetase. Those enzymes that catalyse such syntheses and do not use the energy of a phosphoanhydride bond are synthases (despite their frequently being misnamed).

synthetic peptide combinatorial library (SPCL) A strategy for determination of the specificity of a soluble peptide. Random collections of all the possible oligopeptides of a certain length, preceded by all the possible dipeptide sequences, are prepared; for example, if hexamers, there will be 400 (20^2) sets, each with the two known N-terminal residues followed by a random tetrapeptide; each of the 400 sets will consist of 160 000 (20^4) hexapeptides. Each of the 400 is tested in an assay or bioassay, e.g. binding, inhibition, vasoconstriction. The successful set is refined by placing at the first of the random positions each of the amino acids, so there will be 20 sets of hexapeptides, each with a known tripeptide sequence followed by random tripeptide sequences. These are again tested in the assay system to further narrow the selection. Eventually the most favoured hexapeptide sequence is identified. A related strategy is *positional scanning SPCL*, in which a single amino acid is placed in each of the positions of the oligopeptide, a hexapeptide in the example, with each of the other five positions being filled at random by other amino acids. Each of these sub-libraries is assayed for selection and further refinement as in the original strategy.

- Houghten, R.A. (1994) Methods Companion Methods Enzymol. **6**, 354–360

systematic evolution of ligands by exponential enrichment (*see* cyclic amplification and selection of targets)

T

TAD Transactivation domain. (*see trans-acting*)

tagged random PCR A method for representatively sampling and amplifying DNA. The first two PCR cycles use primers of random sequence, and subsequent cycles use specific primers.

tailing A procedure that is part of one method for insertion of a polynucleotide into a circular plasmid. Addition of a homopolymer to the 3′-end of one unit and the complementary homopolymer to the 3′-end of the other anticipates the annealing of these homopolymers and filling in of the gaps in the new duplex by a polymerase.

tailward (*see headward*)

tandem array A DNA structure in which a gene and associated sequences are repeated in an immediately adjacent position.

tandem enzymes A pair of opposing enzymic activities that are resident in the same polypeptide chain, e.g. isocitrate dehydrogenase kinase and phosphatase; phosphofructokinase 2 and fructose bisphosphatase 2.

tandem mass spectrometry (tandem MS; MS/MS) An adaptation of MS for the analysis of molecular structure; an initial separation by MS generates a peak, i.e. a molecular ion or large fragment, that is further fragmented by gas-phase collision, the daughter fragments of which themselves separated by MS. When more than two stages are involved, the technique is called *multi-dimensional MS* (MS^n, where *n* indicates the number of stages).

- Grant, R.R. and Cooks, R.G. (1990) Science **250**, 61–68

tandem MS (= tandem mass spectrometry)

tannin A product of plant origin that converts hides into leather; often polyphenols and aldehydes that cross-link collagen chains and render them insoluble and unreactive to enzymes that would degrade and putrefy untreated hides. Chromic acid salts are 'tanning agents' but not tannins.

***Taq* polymerase** *Thermus aquaticus* polymerase. (*see polymerase chain reaction*)

target cell A cell that is stimulated by a particular hormone or neurotransmitter.

target detection assay (*see cyclic amplification and selection of targets*)

target organ An organ that is stimulated by a particular hormone.

TATA box A consensus sequence of eukaryotic DNA that is part of the promoter region, and defines the transcription start site downstream from it; also known as a Hogness box or a Pribnow box.

taut conformation (*see Monod–Wyman–Changeux model*)

tautomer An alternative arrangement of the chemical bonds of a molecule that requires movement of only electrons and protons, e.g. the enol form of a carbonyl.

taxon A taxonomic group, such as a family or species.

TCA cycle (= tricarboxylic acid cycle)

T-cell Thymus lymphocyte, which recognizes foreign proteins on the surface of other cells. The groups of T-cells include *cytotoxic T-cells* (*CTLs; killer T-cells*), which can kill the foreign cell; *helper T-cells*, which bind to and assist the proliferation of B-cells; and *suppressor T-cells*, which attenuate the action of B-cells. (*see also* anti-ergotypic; natural killer cell)

TDA Target detection assay. (*see* cyclic amplification and selection of targets)

TEF (= thermic effect of food)

teichoic acid A linear polymer present in Gram-positive bacterial capsules, cell walls and membranes, characterized by polyols (glycerol, ribitol, sugars), sometimes derivatized as D-alanine esters, and linked together through phosphodiester bonds.

telecrine function The action of a hormone distant from its site of secretion. (*see also* paracrine)

temperate phage A virus that infects a bacterium and inserts its genome into that of the bacterium, rather than lysing it.

temperature jump (*see* relaxation)

template A polynucleotide that encodes the information from which another polynucleotide, of complementary sequence, is synthesized.

tense form (= taut conformation; *see* Monod–Wyman–Changeux model)

ter In enzyme kinetics, a designation of a reaction with four substrates or products; in genetics, a designation of the end of a chromosome. (*see also* centrosome; enzyme mechanism)

terminal glycosylation The final modifications of the N-linked carbohydrate moieties of a protein that are effected in the Golgi apparatus. (*see also* core glycosylation)

termination factor A protein that assists in the termination of the action of an RNA polymerase, e.g. the rho factor.

terminator codon A polynucleotide triplet that signals the limit, at the C-terminus, of protein synthesis.

terpene A compound synthesized from isoprene units (biosynthetically, isopentenylpyrophosphate) and recognized in products as a branched five-carbon motif. A *monoterpene* is composed of two isoprene units; a *sesquiterpene* of three; a *diterpene* of four; and a *triterpene* of six.

tertiary structure The unique three-dimensional structure of a particular protein or nucleic acid. (*see also* primary structure; quaternary structure; secondary structure; supersecondary structure)

tester (*see* subtractive DNA cloning)

tetra-antennary Descriptive of a complex-type carbohydrate that has four unbranched oligosaccharide chains at a non-reducing end, often terminated by a sialic acid. Variants are *tri-antennary* and *bi-antennary* structures.

tetrad A four-stranded DNA structure, e.g. the i-tetrad, composed of parallel and anti-parallel poly(C) sequences in which all strands are equivalent and each C–C base pair shares a proton.

• Gehring, K., Leroy, J.-L. and Gueron, M. (1993) Nature (London) **363**, 561–565

tetrahedral intermediate A proposed transition state in the hydrolysis of a peptide or ester bond in which a nucleophile, water or a functional group of the enzyme catalytic site has added across the carbonyl group.

T-form Tense or taut form. (*see* Monod–Wyman–Changeux model)

thanatogene (*see* oncogene)

thermic effect of food (TEF) The energy expended on digestion of food. It is also known as specific dynamic action.

thermodynamic model of protein structure (*see* framework model)

thermogenesis The process of generation of heat by the uncoupling of electron transport, especially in brown adipose tissue.

thermophile (*see* extremophile)

thick filament In muscle cells, an aggregate of myosin molecules in which the long, fibrous tails are intertwined in a superhelix that leaves the globular heads as knobs in a helical array over the surface. The myosin molecules are oriented so that their tails are directed towards the centre of the filament and consequently leave this region of the filaments bare of the knobby projections.

thin filament In muscle and other cells, polymerized actin and associated proteins

thin-layer chromatography (TLC) A technique to separate small molecules on a thin film of an adsorbent on a glass or semi-rigid plastic surface by their differential mobilities in a liquid phase that passes through the film by capillary action.

thiocyanate degradation A method for analysis of peptide structure by repetitive sequential degradation from the C-terminus. The free α-carboxyl group is activated with acetic anhydride and reacted with thiocyanate, and the terminal residue is then hydrolysed off as a thiohydantoin. (*see also* Edman degradation)

thiol proteinase (= cysteine proteinase)

thiolysis In β-oxidation, cleavage of the bond between the β- and γ-carbons by coenzyme A to form acetyl-coenzyme A from carbons α and β and a coenzyme A thioester from the alkyl-carbonyl moiety, i.e. carbons originally γ, δ, ...

thiostatin (T-kininogen) An α_1-cysteine proteinase inhibitor and kininogen of the rat.

thiosulphate shunt A pathway for anaerobic metabolism of sulphide by bacteria, e.g. in marine sediments; thiosulphate ($S_2O_3^{2-}$) is an intermediate between sulphide (S^{2-}) and sulphate (SO_4^{2-}) in reduction to the former, oxidation to the latter and disproportionation to both.

• Jorgensen, B.B. (1990) Science **249**, 152–154

thiotemplate mechanism A non-ribosomal, non-RNA-dependent mechanism for the synthesis of bacterial cyclic oligopeptide antibiotics (e.g. gramicidin S) in which the amino acid residues are first transferred from amino acid adenylates to thiol groups of a peptide synthetase that then collects the residues, in sequence, on a phosphopantetheine arm until the mature peptide is cleaved off by a cyclizing transpeptidation reaction.

thromboxane An eicosanoid that features a six-membered cyclic ether nucleus.

thrombus The product of coagulation, composed of cells and proteins, found when blood clots.

thylakoid membrane A membranous structure within chloroplasts that harvests light and synthesizes ATP.

tier The cohort of chains of a branched-chain polysaccharide that all have the same relationship to the branch points. The outer tier of glycogen is all those glucosyl units that are attached in $\alpha 1 \rightarrow 4$ bonds that run from the non-reducing ends to the first $\alpha 1 \rightarrow 6$ branch points; the second tier is those glycosyl units between the first and second outer branch points, etc. (*see also* limit dextrin)

tight junction The seal that closes the gap between adjacent epithelial cells to ensure closure of a vascular space. Transport through a tight junction is *paracellular*, in contrast to *transcellular* transport through cells.

time constant In first-order kinetics, the reciprocal of the rate constant.

time-of-flight mass spectrometry (TOF-MS) A variant of MS in which the mass of a molecular ion is calculated from the time of travel from the point of ionization to the detector. TOF-MS uses a pulsed ionizing source such as a laser or ^{252}Cf decay into one daughter isotope that ionizes the test material and a second that signals the beginning of the pulse. (*see also* mass spectrometry)

time-resolved crystallography (= Laue crystallography)

TIMP (= tissue inhibitor of metalloproteinase)

tissue A group of cells, of one or several types, that serve a specific function, e.g. adipose tissue, epithelium.

tissue engineering A group of procedures for replacement of damaged or non-functional cells or organs with homologously or heterologously transplanted cells.

• Langer, R. and Vaccanti, J.P. (1993) Science **260**, 920–926

tissue inhibitor of metalloproteinse TIMP; matrixin; a natural inhibitor of the metalloproteinases that remodel structural proteins of connective tissues.

titin An exceptionally large muscle protein, 3000 kDa in molecular mass and 1 μm in length, that spans the entire length of a sarcomere.

titration curve A graphical representation of the protonic dissociation of a compound, e.g. pH against equivalents of alkali added to the acid form of the compound.

T-jump Temperature jump. (*see* relaxation)

TLC (= thin-layer chromatography)

T-lymphocyte (= T-cell)

TOF-MS (= time-of-flight mass spectrometry)

tolerance The property of the immune system that prevents development of autoimmune responses; the suppression of lymphocytes that would otherwise produce autoantibodies.

topoisomers Double-stranded DNA molecules that differ only by their linking numbers.

topology The nature of the supercoiling of a double-stranded DNA molecule; also, in protein chemistry, a formalized array of secondary structures within a molecule. (*see also* Greek key)

• Orengo, C.A., Jones, D.T. and Thornton, J.M. (1994) Nature (London) **372**, 631–634; Efimov, A.V. (1994) FEBS Lett. **355**, 213–219

topology/packing diagram A schematic representation of secondary structure of a protein in which α-helix is rendered as a circle, β-sheet as a rectangle and connecting sequences as lines. The circles and rectangles are arranged in two dimensions to approximate their true orientation in space. Proteins fall into four classes: all-α, which have only helix secondary structure; all-β, which have only sheet; α + β, which have regions of both helix and sheet; and α/β, which have approximately alternating helix and sheet. (*see also* Rossmann fold)

toroidal (*see* plectonemic)

torsional stress (*see* entropy effect)

totipotency The capacity of a zygote to develop into a fully differentiated organism, a property lost as the zygote develops into more specialized cells that are more committed to particular lines of development.

T phage (*see* bacteriophage)

TPN⁺ Triphosphopyridine nucleotide; obsolete name for NADP⁺.

TPNH Reduced triphosphopyridine nucleotide; obsolete name for NADPH.

trace element A constituent of a tissue, a cell or the diet that is present in very low amounts, e.g. copper, zinc, molybdenum.

tracer A compound that is labelled by a radioactive or non-radioactive isotope to assist in the study of its transport or metabolism.

traffic ATPase A class of transmembrane transporters that includes the cystic fibrosis transmembrane regulator (CFTR) and transporters responsible for multidrug resistance. Transport is coupled to ATP hydrolysis but not to the counter- or co-transport of any other metabolite.

trafficking The transport of substances, e.g. a newly synthesized or modified nucleic acid or protein, from one intracellular

structure to another. Intracellular transport of proteins is known as protein kinesis.

trancytosis (*see* caveolae)

trans (*see* cis)

trans-acting Descriptive of a controlling effect of a regulatory gene on a structural gene at some distance from it on the same chromosome or on a different chromosome. The terminology was later broadened to distinguish intermolecular (*trans*) from intramolecular (*cis*) actions. A regulatory factor mediates binding of activators and repressors via its DNA-binding domain and its transactivation domain. At higher levels, the concentration of the factor exceeds that of its DNA sites and, in solution, it can bind the activator or repressor so as to preclude binding of the latter to any DNA-bound factor, and can thus prevent or 'squelch' activation or repression. (*see also* cis-acting)

• Cahill, M.A., Ernst, W.H., Janknecht, R. and Nordheim, A. (1994) FEBS Lett. **344,** 105–108

transamidation The substitution of one amino compound for another in an amide linkage, e.g. the cross-linking of fibrin by the displacement of NH_3 from glutamine residues by an ε-amino group of a lysine residue to form an isopeptide bond.

transamination The concurrent amination and reduction of a carbonyl of one compound as another is deaminated and oxidized; especially the conversion of one α-amino acid into its corresponding α-oxo acid as another α-oxo acid is converted into its corresponding α-amino acid, mediated by the pyridoxal phosphate prosthetic group of a transaminase (aminotransferase). (*see also* quinimine form)

transcribed strand (*see* antisense)

transcription Synthesis of RNA from a DNA template.

transcriptional activation The process of separation of strands of DNA at which replication will commence. Short RNA sequences hold apart the DNA strands to allow a primosome to bind and synthesize primers of DNA synthesis.

transcriptional silencing The exercise of genetic control to prevent expression of a structural gene.

transcription–translation coupling The immediate use of mRNA transcripts for protein synthesis before the synthesis of the mRNA is complete; occurs in bacteria where RNA processing does not occur.

transcytosis The transport of materials across a polarized cell, e.g. in digestion from the apical surface of a cell of the intestinal epithelium and its movement across the cell to the basolateral surface.

***trans*-dominant** Descriptive of an engineered protein or nucleic acid fragment that inhibits the activity of its parent molecule by preferential binding to the parent's target site.

transduction The conversion of one kind of energy into another, e.g. conversion of the chemical energy of ATP into mechanical energy by muscle contraction; also conversion of a signal as it crosses a barrier, e.g. linkage of hormone binding on the outside of a plasma membrane to the generation of a second messenger inside

it; also DNA transfer by a virus that can incorporate into its own genome part of the DNA of a first host and then transfer it to a second host.

transesterification The substitution of one alcohol for another in an ester bond, e.g. the displacement of one 3′-hydroxy group of a phosphodiester bond by the 3′-hydroxy group of another nucleotide during the self-splicing of an RNA molecule, or the internal transfer of phosphate from a 3′-hydroxy to a 2′-hydroxy group during the reaction of pancreatic ribonuclease.

transfection The process by which viral or bacteriophage DNA is introduced into a cell or bacterium. (*see also* transformation)

transferase One of a class of enzymes that transfer a chemical group from donor substrates to acceptor substrates, e.g. a kinase, a phosphorylase, a transaminase. (*see also* hydrolase)

transfer RNA (= tRNA)

transfer vesicle A portion of the endoplasmic reticulum which encloses newly synthesized proteins, that buds off and travels to and fuses with the Golgi apparatus; also those vesicles that transfer protein from the Golgi apparatus to lysosomes, secretory vesicles and the plasma membrane.

transformant A product of transformation; a transformed cell.

transformation The process by which a cell line, that can normally be expected to undergo a limited number of cell divisions before death, becomes immortal; also the process by which isolated foreign DNA is introduced into a cell or bacterium.

transforming factor In the classical experiments by Avery, MacLeod and McCarty, the name given to the agent that caused the heritable transformation of R strains of *Streptococcus pneumoniae* into S strains. The identification of the factor as DNA was the definitive proof that DNA is the repository of genetic information. (*see also* lipopolysaccharide)

***trans*-fusion** (*see* alpha configuration)

transgene DNA that has been experimentally introduced into a transgenic animal.

transgenic Descriptive of an animal that has developed in a surrogate mother from a fertilized ovum that had been injected with a recombinant DNA gene; contrasted with *congenic*, meaning of the same genetic origin.

***trans*-Golgi** (*see* Golgi apparatus)

transition In replication or transcription, an error in which one purine is substituted for another, or one pyrimidine for another. (*see also* transversion)

transition state In chemical kinetics, the hypothetical state that is mid-way between reactants and products, poised at a point where the reaction is as likely to go forward to products as it is to fall back to reactants. If the reactants and transition state may be considered to be in equilibrium, application of the second law of thermodynamics allows analysis of the temperature-dependence of reaction rates. The concept may be applied to enzyme kinetics to describe the transition from enzyme–substrate complex to enzyme–product complex.

transition state inhibitor An enzyme inhibitor that is designed to fit the transition state of an enzyme, as opposed to one that is a substrate analogue; characterized by a very low dissociation constant.

transition temperature The temperature at which the plasma membrane undergoes a phase transition, due to the increased mobility at higher temperatures of the fatty acyl chains of phospholipids.

translation The synthesis of a protein directed by mRNA.

translocation In genetics, the movement of a portion of one chromosome to another; in protein synthesis, the transfer of the newly elongated peptidyl-tRNA from the amino acyl site to the peptide site of a ribosome; in cell biology, the movement of a molecule across a barrier or between cytosol and membrane surface.

transmembrane domain A feature of most intrinsic proteins of plasma or vesicular membranes; a polypeptide sequence of about seven residues if β-sheet, up to 22 residues if α-helix, that connects extracellular to intracellular domains, joined by extended polypeptides on the cytoplasmic and external or vesicular sides. Receptor and ion channels have between one and twelve such domains.

transmethylation The metabolic transfer of preformed methyl groups from one acceptor to another, e.g. from *S*-adenosylmethionine to guanidinoacetate to form creatine.

transparent mutation A single base change that does not result in an altered protein, due to the degeneracy of the genetic code, i.e. the mutation is from one codon to another for the same amino acid. (*see also* silent mutation)

transpeptidation A characteristic of some proteinases; the transfer of one product of the catalytic cleavage to another peptide rather than to water; the transfer of the N-terminal fragment to the amino group of an acceptor peptide or the transfer of the C-terminal fragment to the carboxyl group of an acceptor; the former often, but not exclusively, due to an acyl-enzyme intermediate in the catalytic cycle.

transposable element (= transposon)

transposition The movement of a mobile element into or out of a chromosome or plasmid.

transposon One of a class of genes that are capable of moving spontaneously from one chromosome to another, or from one position to another in the same chromosome; also known as jumping genes or transposable elements.

transposon tagging The use of a transposon to isolate a gene whose product may be unknown. Mutants in which a transposon inserts into and disrupts the function of a gene are selected by phenotype. A DNA fragment that contains the gene with its insertion is recognizable and is selected by its binding of a probe complementary to the transposon insert. The DNA whose sequence flanks the insert is isolated and used as a probe to screen a library of non-mutants for the intact gene.

• Watson, J.D., Gilman, M., Witkowski, J. and Zoller, M. (1992) Recombinant DNA, 2nd edn., pp. 186–188, Scientific American Books, New York

trans-**recognition** A refinement of the N-end rule; the determination of the rate of degradation of a cellular multimeric protein composed of different kinds of subunits by amino acid residues of a subunit other than the one that is rapidly degraded. *cis*-Recognition is the targeting by residues of the subunit that is rapidly degraded. The distinction implicitly acknowledges that such subunits may be degraded at different rates.

• Johnson, E.S., Gonda, K. and Varshavsky, A. (1990) Nature (London) **346**, 287–291

transverse diffusion (= flip-flop)

transversion In replication or transcription, an error in which a purine is substituted for a pyrimidine, or a pyrimidine for a purine. (*see also* transition)

transwinding The action of a protein to separate one RNA strand from its hybrid with a second, and to promote its re-annealing with a third RNA strand.

trefoil (*see* node)

tri-antennary (*see* tetra-antennary)

tricarboxylic acid cycle (TCA cycle) The metabolic pathway in which acetyl-CoA is catalytically oxidized to carbon dioxide, with the concomitant reduction of NAD^+ and FAD via a series of tricarboxylic (citric, cisaconitic and isocitric) and dicarboxylic (succinic, fumaric, malic and oxaloacetic) acids. The actual carbon atoms that appear as CO_2 after a single passage through the cycle are not identical to those that entered as acetyl-CoA. Also known as the citric acid cycle or the Krebs cycle.

trigonal bipyramid (*see* pentacovalent intermediate)

trinucleotide repeat A mutation in which a trinucleotide is repeated, even thousands of times; a characteristic of several diseases, e.g. fragile X syndrome.

triple helix The tertiary structure of collagen that twists three polypeptide chains around themselves; also a triple-stranded DNA structure that involves Hoogstein base pairing between B-DNA and a third DNA strand that occupies the major groove.

triplet A three-base codon of the genetic code. Also a component of the mitotic spindle, seen in cross-section as one of nine structures, each composed of a microtubule doublet with an additional series of protofilaments that constitute a third subunit; the nine triplets radiate from a central axis and appear as a *cartwheel*.

triplet state An excited electronic state in which an electron is raised to a higher energy level and its spin is reversed, which prevents its easy collapse to the ground state. Such collapse is sometimes accompanied by phosphorescence. (*see also* singlet state)

triplex DNA Triple-stranded DNA in which the third oligonucleotide strand lies in the major groove of duplex DNA. It may also be a synthetic oligonucleotide with a sequence designed to target a specific sequence of a duplex; it can block transcription or, if it bears a covalently attached reactive group, it can cleave the duplex at a unique site.

- Moffat, A.S. (1991) Science **252**, 1374–1375

triskelion The three-legged structure of clathrin, composed of three heavy and three light chains. Subunits of the polyhedral shell surround coated vesicles.

triterpene (*see* terpene)

tRNA Transfer RNA; the RNA that serves in protein synthesis as an interface between mRNA and amino acids. It carries an anticodon sequence that pairs bases with a codon of mRNA, and it binds an amino acid at its 3′-end through an ester bond.

Trojan horse inhibitor Affinity labelling reagent.

trophic Descriptive of the dependence of a cell, tissue or organism on an external agent for its maintenance or growth. (*see also* tropic)

tropic Descriptive of a response of a cell, tissue or organism to a stimulus, especially a hormone. (*see also* trophic)

trp operon A region of DNA that, when expressed, results in an RNA transcript that codes for five enzymes that act sequentially on chorismic acid to transform it into tryptophan.

truncation Elimination of the N- or C-terminal portion of a protein by proteolysis or manipulation of the structural gene, or premature termination of protein elongation due to the presence of a termination codon in its structural gene as a result of a nonsense mutation.

ts A prefix that indicates a temperature-sensitive form.

T series phage One of a group of bacteriophage that infect *E. coli*; the subject of extensive experimentation.

tumbling The irregular motion of a flagellar bacterium in solution. (*see also* random walk)

tumour A neoplasm; a growth of new tissue, sometimes malignant.

tumour initiator An agent that damages cellular DNA, a necessary condition for tumorigenesis. (*see also* anti-oncogene; tumour promoter)

tumour promoter An agent that converts a cell with damaged DNA into a malignant cell. (*see also* anti-oncogene; tumour initiator)

tumour suppressor gene (= anti-oncogene)

turgor The internal hydrostatic pressure of a cell, especially a plant cell.

turnover Degradation and resynthesis of a compound or macromolecule.

turnover number (catalytic-centre activity) The number of substrate molecules that a single catalytic site of an enzyme molecule can convert into product in a given time. (*see also* catalytic constant)

twintron An intron within an intron; the internal intron is excised before the external intron is spliced out.

• Copertino, D.W. and Hallick, R.B. (1993) Trends Biochem. Sci. **18**, 467–471

twist conformation A conformation of a five-membered ring in which no four centres lie in the same plane. (*see also* chair form; envelope conformation)

twisting number (*T*) A topological property of double-stranded DNA; the number of turns one DNA strand makes around the other; the number of paired bases in a DNA molecule divided by 10; equal to the linking number minus the writhing number ($T = L - W$).

two-component pathway A pattern of molecular organization by which responses to external stimuli are processed by cells to effect modulation of output, such as chemotaxis in bacteria, or osmolarity regulation in plant cells. An example is a kinase that, following its interaction with an activated receptor, phosphorylates a function-specific response regulator.

• Koshland, D.E., Jr. (1993) Science **262**, 532

two-dimensional DNA typing Characterization of DNA according to base composition by electrophoresis, first according to chain length in one direction, then at 90° in a denaturant (e.g. formamide, urea).

• Mullaert, E., de Vos, G.J., teMeerman, G.J., et al. (1993) Nature (London) **365**, 469–471

two-dimensional electrophoresis Electrophoretic separation of proteins on a solid support by one technique in one dimension, then by another technique in another. For example, proteins are subjected to isoelectric focusing in a long, thin tubular gel, then the developed gel is placed at the top of a sodium dodecyl sulphate/polyacrylamide slab and subjected to gel electrophoresis in the other direction.

two-dimensional homochromatography A form of displacement chromatography in which, after an initial separation in the first dimension of a labelled oligomer (e.g. an oligoribonucleotide), separation in the second dimension proceeds in the presence of a high concentration of a heterogeneous mixture of oligomers (e.g. an unfractionated endonuclease digest of yeast RNA).

two-hybrid system A strategy for identification of protein–protein interactions by utilizing a eukaryotic expression system such as yeast cells. Two hybridized proteins, products of genetic engineering, are created and expressed in the test system: one which contains one of the proteins to be tested for interaction fused to a domain that binds to DNA upstream from a reporter gene (e.g. one that encodes an enzyme that hydrolyses a chromogenic substrate); the other hybrid contains the second protein to be tested for interaction, fused to a transcriptional activation domain. In screening these hybridized molecules, the appearance of the coloured product of the enzyme encoded by the reporter gene signals the interaction of the pair of proteins that are moieties of the hybrids. In the *one-hybrid* variation, fusion to a transcription-activating domain may similarly detect a DNA-binding protein.

- Allen, J.B., Walberg, M.W., Edwards, M.C. and Elledge, S.J. (1995) Trends Biochem. Sci. **20**, 511–516

two-state rule In protein chemistry, the postulate that the folding of a protein occurs without any stable intermediate between the denatured and native states. (*see also* framework model; molten globule model)

- Baldwin, R.L. (1990) Nature (London) **346**, 409–410

Tyndall effect The phenomenon of light scattering that is dependent upon the wavelength of light. The intensity of light scattered by small isotropic particles is inversely proportional to the fourth power of the wavelength.

tyrosine kinase An enzymic activity associated with the cytoplasmic domain of several growth factor receptors. Two such molecules, when liganded to an effector molecule by their extracellular domains, phosphorylate tyrosine residues on each other; this increases their kinase activities so that they are then able to phosphorylate other cytoplasmic proteins, thus continuing the signal transduction pathway initiated by effector binding to the receptor.

tyrphostin An inhibitor of a tyrosine kinase.

U

ubiquitin pathway A route to cellular proteolysis that depends upon conjugation of the targeted protein to a small protein, ubiquitin, followed by hydrolysis.

• Ciechanover, A. and Schwartz, A.L. (1994) FASEB J. **8**, 182–191

U-DNA (*see* Kunkel mutagenesis)

ultracentrifugation (*see* analytical ultracentrifugation; preparative ultracentrifugation)

ultrafiltration A separation procedure in which a solution is forced through a membrane with a pore size that is selected to retain macromolecules of a certain size and to pass smaller ones.

unassigned reading frame (URF) An open reading frame for which a protein product has not yet been identified.

uncompetitive inhibition A form of enzyme inhibition in which the inhibitor binds to the enzyme–substrate complex, resulting in decreases in the K_m and V_{max} values. (*see also* competitive inhibition; inhibitor; non-competitive inhibition)

uncoupling agent A compound that dissociates electron transport from ATP synthesis and allows transport to proceed without synthesis. (*see also* oxidative phosphorylation)

uni (*see* enzyme mechanism)

uniport A transport mechanism that drives a single compound or ion across a membrane, not coupled with transport of any other compound or ion. (*see also* antiport; mobile barrier; mobile carrier; symport)

unit In enzymology, a measure of enzyme activity; usually the conversion of 1 μmol of substrate per min under specified conditions. (*see also* katal)

unit cell The basic repeating unit of a crystal.

unit evolutionary period A measure of the rate of divergence of two daughter genes after duplication of the ancestral gene; the time taken to accumulate 1% divergence in primary structures.

unit membrane Danielli–Davson model A model for the structure of biological membranes; a variation of the Danielli–Davson model in which the membrane is composed of plaques of bilayer (the unit membrane); the outer leaflet turns over the edges of the plaque to become continuous with the inner leaflet. (*see also* fluid mosaic model; Gortner and Grendel model)

universal In molecular biology, descriptive of a tool that is generally useful, not limited to a specific DNA fragment or sequence; e.g. a universal adaptor is one that can be ligated to all DNA fragments.

unnatural amino acid mutagenesis A method for replacement of an amino acid residue of a protein with a non-naturally occurring amino acid. The mRNA that encodes the protein is modified to encode a nonsense suppressor codon. For example, mRNA is modified to use the UAG codon; this mRNA, together with a nonsense suppressor tRNA with the anticodon CUA, and with an unnatural amino acid at its 3′-end, is introduced into an expression system, possibly a *Xenopus* oocyte.

• Nowak, M.W., Kearney, P.C., Sampson, J.R., et al. (1995) Science **268**, 439–442

unscheduled DNA synthesis Replication that occurs outside the synthetic (S) phase of the cell cycle.

untranslated region (UTR) A genomic DNA sequence that is not translated into an RNA sequence.

upfield Descriptive of a resonance position in a magnetic field higher than that at which a standard displays its signal.

upstream In a polynucleotide chain, towards the 5′-end. (*see also* downstream)

urea cycle (Krebs urea cycle) The metabolic pathway that receives nitrogen and carbon dioxide, as carbamoyl phosphate and aspartic acid, and synthesizes urea.

ureotelic Descriptive of an organism that synthesizes and excretes urea as the final product of nitrogen metabolism. (*see also* ammonotelic; uricotelic)

URF (= unassigned reading frame)

uricotelic Descriptive of an organism that synthesizes and excretes uric acid as the final product of nitrogen metabolism. (*see also* ammonotelic; ureotelic)

uronic acid A sugar derivative in which the hydroxymethyl group is oxidized to a carboxyl group.

UTR (= untranslated region)

V

v A prefix that denotes viral. (*see also* onco-gene)

vacuole A large vesicle, especially in a plant cell.

van den Bergh reaction A method for colorimetric estimation of serum bilirubin by coupling with diazotized sulphanilic acid. The *direct van den Bergh reaction*, in aqueous medium, yields the 'direct' bilirubin, i.e. the amount of water-soluble bilirubin glucuronide conjugates. Upon prior addition of methanol to solubilize free bilirubin, the total bilirubin is found, and by this *indirect van den Bergh reaction* the 'indirect' bilirubin, i.e. the amount of unconjugated bilirubin, may be calculated as the difference between the total and direct levels.

van der Waals bond The weak attraction of neighbouring neutral atoms that includes dipole–dipole and dipole-induced dipole interactions and London dispersion forces.

van der Waals radius The dimensions of an atom deduced from its packing in crystals.

Van Slyke method A method to quantify primary amino groups by measurement of the volume of nitrogen produced upon reaction with nitrous acid; an obsolete method for the assay of proteinase action.

variable number tandem repeat (VNTR) A minisatellite DNA; the repetition of a 35–80 bp sequence, to a size of up to 2 kbp, that is characteristic of an individual. VNTRs are used in forensic science to identify, or exclude, suspects of a crime. (*see also* DNA fingerprinting)

variable region The sequences of immunoglobulin light or heavy chains that show variation from one particular antibody to another, and are responsible for the binding of antigen. (*see also* constant region; hypervariable region)

vasoconstriction The decrease in diameter of small blood vessels, especially in the kidney, that raises blood pressure. (*see also* vasodilation)

vasodilation The increase in diameter of small blood vessels, especially in the kidney, that lowers blood pressure. (*see also* vasoconstriction)

VCD Vibrational circular dichroism. (*see* vibrational optical activity)

vector A DNA molecule that can be replicated in a cell and that can serve as the vehicle for transfer to such a cell of DNA that has been inserted into it by recombinant techniques.

vectorette PCR A PCR variation that permits specific amplification of the sequence from a single known internal sequence to the end of the fragment. The source DNA is cleaved by a restriction endonuclease that leaves an overhang, by which the fragments are ligated to a compatible universal synthetic oligonucleotide duplex called a vectorette. The vectorette is engineered with a central mismatched sequence, and the vectorette primer is identical to the sequence on the vectorette strand that becomes ligated to that fragment strand which can bind a second primer that is complementary to the known internal sequence of the fragment. The binding site for the vectorette primer is created only when the internal primer is extended to the end of the vectorette. This avoids duplication of the vectorettes attached to all other restriction fragments. A refinement is *splinkerette PCR*, in which the primers are designed to form hairpin structures, which decreases the danger of non-specific priming. (*see also* inverse PCR)

• Hengen, P.N. (1995) Trends Biochem. Sci. **20**, 372–373

Velcro mechanism (cysteine switch) An activation mechanism for matrix metalloproteinase family zymogens (named in reference to the zipper-type mechanism), in which the thiol group of a cysteine residue in the N-terminal propart binds to and blocks the active-site Zn atom so that activation occurs when the thiol group is itself blocked, e.g. by *N*-ethylmaleimide or a heavy metal, or when the propart is proteolytically excised. (*see also* peptide Velcro)

• Vallee, B.L. and Auld, D.S. (1990) Biochemistry **29**, 5647–5659; Van Wart, H.E. and Birkedal-Hansen, H. (1990) Proc. Natl. Acad. Sci. U.S.A. **87**, 5578–5582

very-long-chain fatty acid (*see* medium-chain fatty acid)

vesicle A membrane-limited sac in the cytoplasm of a cell.

vesicular traffic The transport of intermediates and finished products from intracellular membrane sites to other intracellular or plasma membrane sites.

V_H The variable domain of an immunoglobulin heavy chain. (*see also* V_L)

vibrational circular dichroism (*see* vibrational optical activity)

vibrational optical activity (VOA) A spectroscopic technique for examination of chiral structure, especially of biopolymers, in which the vibrational spectra (infrared and visible), due to bond stretching and bending [as opposed to electronic spectra (ultraviolet), due to electron excitation], of chiral centres are isolated from the background spectra. *Raman optical activity* (*ROA*) observes the difference in intensities of right- and left-circularly polarized light, and is conventionally displayed as a plot of the difference in intensities, I^R - I^L (arbitrary units), against wavenumber (cm^{-1}). *Vibrational circular dichroism* (*VCD*) is essentially CD in the infrared spectrum. Analysis of data is by *ab initio* calculation for small molecules, or by empirical correlation with VOA of known structures for larger molecules. (*see also* light scattering)

• Freemantle, M. (1995) Chem. Eng. News **73 (3)**, 27–34; Carey, P.R. and Tonge, P.J. (1995) Acc. Chem. Res. **28**, 8–13

vicinal 'Neighbouring'; descriptive of substituents of adjacent carbon atoms, e.g. vicinal hydroxy groups of a sugar.

vinca alkaloid Also called vinblastine or vincristine; an inhibitor of microtubule polymerization and an anti-mitotic agent.

virion The extracellular form of a virus, complete with its capsid.

viroid A small circular single-stranded RNA that can infect plants.

virus A primitive life form; a nucleic acid surrounded by a protein coat that requires for its own propagation the infection of a plant or animal cell or a bacterium; viruses include important pathogens and useful agents for the study of cellular processes.

visual cascade The sequence of enzymic and non-enzymic events that is triggered by absorption of light in the retina and results in transmission of a nerve impulse.

vitamin A compound required in the diet in minute amounts; fat-soluble if soluble in organic solvents (vitamins A, D and E); water-soluble if soluble in aqueous solutions (ascorbic acid, biotin, cobalamin, folic acid, nicotinamide, pantothenic acid, pyridoxine, riboflavin and thiamin).

V_L The variable domain of an immunoglobulin light chain. (*see also* V_H chain)

V_{max} (= maximum velocity)

VNTR (= variable number tandem repeat)

VOA (= vibrational optical activity)

W

Walden inversion A change of the configuration at an asymmetrical carbon atom due to the entry from one side of the centre of an attacking group, simultaneously with the departure from the other side of a leaving group.

Warburg apparatus (*see* manometry)

Warburg–Dickens pathway (= pentose phosphate pathway)

Watson–Crick base pairs The base pairs that are compatible with a DNA double helix; i.e. adenine with thymine, and guanine with cytosine.

Watson–Crick model (= double helix)

Watson strand (*see* Crick strand)

wax The product of esterification of a fatty acid with an alcohol other than glycerol, usually a long-chain alcohol.

weaning The removal of a young mammal from a diet of its mother's milk.

weanling A newly weaned animal.

western blotting (immunoblotting) A technique to detect specific proteins. A mixture of proteins is separated by polyacrylamide-gel electrophoresis, blotted on to a plastic sheet (e.g. of nitrocellulose) and then exposed to a radiolabelled immunoglobulin directed against the desired protein, which is revealed by autoradiography. (*see also* electroblotting; northern blotting; Southern blotting; south-western blotting)

white muscle (fast-twitch muscle) A poorly vascularized form of skeletal muscle characterized by little myoglobin and few mitochondria; supplied with energy mainly by glycolysis. (*see also* red muscle)

wild type In genetics, the unmutated allele or the natural allele that predominates over other variants in a population.

wobble The presumed slackness of structural requirements for complementarity in fitting codon to anticodon that permits several bases in the third position of the codon to pair with a particular base in the first position of the anticodon.

Wood–Werkman reaction The proposed reaction, derived from the observed incorporation of isotopic CO_2 into succinate, by which CO_2 condenses with the 3-carbon compound pyruvate to form a 4-carbon compound, oxaloacetate. The basis of the observation was later recognized as the pyruvate carboxylase reaction.

writhe (*see* node)

writhing number (*W*) A topological property of double-stranded DNA; the number of times one double-stranded segment crosses another in circular DNA, commonly referred to as a supercoil; equal to the linking number minus the twisting number ($W = L - T$). (*see also* node)

X

xaptonuon (*see* junk DNA)

xenobiotic A non-biological compound, often one that an organism must eliminate or neutralize by some detoxification strategy.

***Xenopus* oocyte** An egg of the African clawed frog, *Xenopus laevis*; a favoured heterologous expression system that, because of its large size, can be micro-injected with cDNA or mRNA and monitored for synthesis of a specific protein.

X-ray crystallography A method for structural analysis of solids that have a repeating unit; dependent upon the ability of electrons to scatter X-rays, and capable of locating in space atoms larger than hydrogen in proteins and nucleic acids as well as in less complex compounds. (*see also* neutron diffraction)

Y

YAC (= yeast artificial chromosome)

Y-DNA (= Y-joint)

yeast artificial chromosome (YAC) A cloning vector that can accept a relatively large fragment of foreign DNA, up to 1 Mb, in yeast cells. (*see also* bacterial artificial chromosome; cosmid)

- Watson, J.D., Gilman, M., Witkowski, J. and Zoller, M. (1992) Recombinant DNA, 2nd edn., pp. 590–592, Scientific American Books, New York

yellow enzyme One of the two first-discovered flavoenzymes. The 'new' yellow enzyme (das neue gelbe Ferment) catalyses the reduction of molecular oxygen or Methylene Blue by glucose 6-phosphate, as is now understood via reduction by NADPH, and has an FAD prosthetic group. The first-discovered 'old' yellow enzyme (*das alte gelbe Ferment*) has an FMN prosthetic group but has no recognized biological function.

- Karplus, P.A., Fox, K.M. and Massey, V. (1995) FASEB J. **9**, 1518–1526

Yin–Yang hypothesis The now-abandoned theory that bidirectional cellular processes, i.e. those that involve two individual pathways (rather than one that may be 'on' or 'off'), e.g. proliferation/ contact inhibition, glycogen synthesis/ breakdown, are regulated by a balance of the opposed actions of cyclic AMP and cyclic GMP.

Y-joint A DNA heteroduplex structure in which one strand is extended by a non-homologous internal sequence. The loop of the longer strand is base-paired with itself in a hairpin structure. In an *open Y-joint* the shorter, continuous strand is nicked at the joint; in a *closed Y-joint* the continuous strand is un-nicked.

Y-junction (= Y-joint)

Y-structure (= Y-joint)

y-type ion (*see* a-type ion)

Z

Z-DNA A left-handed variant of the DNA double helix originally observed in a sequence of alternating G and C bases, but possible in sequences of alternating purine and pyrimidine bases. The purines are 'flipped' 180° about the glycosidic bond to assume the *syn* conformation (as opposed to B-form DNA, in which they assume the *anti* conformation); the pyrimidines remain in the *anti* conformation. (*see also* A-DNA; B-DNA)

- Herbert, A. and Rich, A. (1996) J. Biol. Chem. **271**, 11595–11598

Zeeman splitting The detail of a spectrum (nuclear magnetic resonance, magnetic circular dichroism, etc.) that is due to the energy differences of spin states in a magnetic field. (*see also* hyperfine splitting)

zero-order kinetics The insensitivity of a rate on the concentration of a reactant; especially for an enzymic reaction at substrate concentrations much above the K_m. (*see also* first-order kinetics)

zeugmatography (spin imaging) Nuclear magnetic resonance imaging of the distribution of a specific compound, e.g. $^{31}PO_4$ species, in a part of the body, for research or medical diagnostics.

zinc finger (metal-binding finger) A motif of which seven or more may appear in a DNA-binding protein. Each is characterized by two closely spaced cysteine and two histidine residues that serve as ligands for a single Zn^{2+}. When bound to Zn^{2+} they form a module from which protrude amino acid side chains that interact with the bases partially exposed to the DNA major groove. Since in the area of contact there is a limited and local parallel of DNA and protein sequences, the occurrence of the contacting amino acid residues in the protein sequence falls into a regularity, or register, that creates in the protein a syllabicity.

- Klug, A. and Schwabe, J.W.R. (1995) FASEB J. **9**, 597–604; Berg, J.M. (1995) Acc. Chem. Res. **28**, 14–19

zinc proteinase A metalloproteinase that features a Zn^{2+}, e.g. carboxypeptidase A, mammalian collagenase.

zipper interaction The sequence of events by which an immunoglobulin G-coated lymphocyte draws the macrophage plasma membrane around it. Fc receptors on the membrane surface attach immunoglobulin G markers and eventually surround the cell and engulf it in a phagosome.

zipper-type mechanism A mechanism for proteolytic degradation of a native protein, in which cleavage of the first peptide bond in each molecule is much faster than subsequent cleavages. The result is that every native molecule has one bond cleaved before any molecule is further affected. In the other alternative extreme case, the *one-by-one type mechanism*, initial cleavage of the first bond of the native molecule is relatively slow but subsequent bond cleavages are fast, which results in complete degradation of one substrate molecule before another is attacked.

Znf (= zinc finger)

zonal centrifugation (= density-gradient centrifugation)

zona pellucida A glycoprotein coat that surrounds an oocyte or ovum.

zonation The phenomenon by which specific areas of an organ have characteristic metabolic functions; e.g. alanine aminotransferase activity is relatively high in the cells of the liver served by the portal blood supply (the periportal region) whereas glutamate dehydrogenase activity is relatively high in the cells that feed the hepatic vein (the perivenous region), and glutamine hydrolysis and urea synthesis are prominent in the periportal region while glutamine synthesis is prominent in the perivenous region.

• Quistorff, B. (1990) Essays Biochem. **25**, 83–136

zone electrophoresis (= free zone electrophoresis; *see* electrophoresis)

zone of adhesion An area of a lipid bilayer where inner and outer leaflets join, and by which proteins, gangliosides and other products may be targeted to the outer surface of the plasma membrane from their cytoplasmic sites of biosynthesis. Models postulate that such zones are continuous, e.g. they surround a pore through the membrane, or occur stepwise via a vesicle formed by the pinching off into the inter-leaflet space of part of the membrane inner leaflet, which can then fuse with the membrane outer leaflet.

• Raetz, C.H.R. (1990) Annu. Rev. Biochem. **59**, 129–170

zoo blotting A technique to assess the degree to which a DNA sequence, usually a coding sequence, is conserved in evolution; a test for hybridization of a probe complementary to the sequence in one species against DNA isolated from cells of a variety of other species.

Z-scheme In the Hill reaction of photosynthesis, the flow of electrons from water through photosystems II and I to reduce NAD^+.

• Prince, R.C. (1996) Trends Biochem. Sci. **21**, 121–122

zwitterion A chemical compound that has positively and negatively charged groups even when it bears no net charge, e.g. glycine, $H_3^+NCH_2COO^-$.

zygote The product of fusion of a male and a female gamete; a fertilized ovum.

zymogen A proenzyme; an inactive precursor of an enzyme, e.g. trypsinogen.

zymogen granule A cytoplasmic vesicle that contains a protein prior to secretion.

zymogram A slab gel that, after electrophoresis, has been stained to reveal the presence of an enzymic activity, e.g. a gel infused or layered with a protein substrate and subsequently treated with a stain for the protein to reveal, as cleared areas, the location of proteolytic activity. (*see also* zymography)

zymography A histological method for localization of an enzymic activity. The tissue or cell preparation overlays a gel that contains a chromophore- or fluorophore-labelled substrate, e.g. casein for a proteinase assay. After enzymic reaction, and washing or diffusing away of labelled products, the sample is coated with a photographic emulsion, exposed and developed to superimpose the histological sample on cleared areas in the labelled substrate that represent areas of enzymic activity. (*see also* zymogram)

zymosan Particles of yeast cell walls.

Zwischenferment Obselete term for the oxidative pathway that generates NADPH from glucose 6-phosphate, now known to be glucose-6-phosphate dehydrogenase plus 6-phosphogluconate dehydrogenase.

ADDITIONAL REFERENCES

Biochemistry

Stenesh, J. (ed.) (1989) Dictionary of Biochemistry and Molecular Biology, 2nd edn., John Wiley & Sons, New York

McGraw-Hill Dictionary of Scientific and Technical Terms, 5th edn. (1994) McGraw-Hill, New York

Biochemical Nomenclature and Related Documents: a Compendium, 2nd edn. (1992), published on behalf of the International Union of Biochemistry and Molecular Biology by Portland Press, London

Lehninger, A.L., Nelson, D.L. and Cox, M.M. (1993) Principles of Biochemistry, 2nd edn., Worth Publishers, New York

Enzyme mechanisms

Cleland, W.W. (1963) Biochim. Biophys. Acta **67**, 104–137: categorization of mechanisms by molecularity

Demuth, H.V. (1990) J. Enzyme Inhib. **3**, 249–278: review of affinity labelling, transition-state analogues and enzyme-activated inhibitors

Enzyme nomenclature

Enzyme Nomenclature Recommendations (1992) of the Nomenclature Committee of the International Union of Biochemistry and Molecular Biology, Academic Press, San Diego

Mass spectrometry

Hellerqvist, C. and Sweetman, B. (1990) Methods Biochem. Anal. **34**, 1–89

Geisow, M. (1991) The Biochemist (Bull. Biochem. Soc.) **13 (3)**, 10–13

Mann, M. and Wilm, M. (1995) Trends Biochem. Sci. **20**, 219–224

Molecular spectroscopy

Wendisch, D.A.W. (1990) Acronyms and Abbreviations in Molecular Spectroscopy: An Encyclopedic Dictionary, Springer-Verlag, Berlin

Glycosaminoglycans

Scott, J.E. (1992) Glycoconjugate J. **10**, 419–421: proposal for designation of polymers by use of a single-letter code.

Molecular biology

Kendrew, J. and Lawrence, E. (1994) The Encyclopaedia of Molecular Biology, 1165 pp., Blackwell Science, Oxford

Ausubel, F.M., Brent, R., Kingston, R.E., et al. (1989) Current Protocols in Molecular Biology (and supplements), pp. 1.0.3–1.0.5, John Wiley & Sons, New York

Genetics and genetic engineering

Stewart, A. (ed.) (1995) Trends in Genetics Genetic Nomenclature Guide, Elsevier Trends Journals, Cambridge: supplement to March 1995 issue of Trends in Genetics, includes information on genomic databases.

Rieger, R., Michalis, A. and Green, M.M. (1991) Glossary of Genetics, 5th edn., Springer-Verlag, Berlin

Oliver, S.G. and Ward, J.M. (1985) A Dictionary of Genetic Engineering, Cambridge University Press, Cambridge

King, R.C. and Stansfield, W.D. (1990) Encyclopedic Dictionary of Genetics, VCH, Weinheim

Biological taxonomy
Roberts, D. (1995) The Biochemist (Bull. Biochem. Soc.) **17(1)**, 25–29

Archaebacteria
Potter, S. (1992) The Biochemist (Bull. Biochem. Soc.) **14(2)**, 21–24

Internet home pages
http:\\www.wehi.edu.au/pedro/research_tools .html (an entrée to internet resources, including dictionaries)
http:\\expasy.hcuge.ch/txt/nomlist.txt (protein nomenclature-related literature references)
http:\\minyos.its.rmit.edu.au/~s9211747/e e.html (capillary electrophoresis)